The demise of a rural economy

International Library of Anthropology
Editor: Adam Kuper, University of Leiden

***Pukhtun Economy and Society**
Traditional Structure and Economic Development in a Tribal Society Akbar S. Ahmed

Power and Independence
Urban Africans' Perception of Social Inequality
P. C. Lloyd

Robber Noblemen
A Study of the Political System of the Sikh Jats
Joyce Pettigrew

Meaning in Culture
F. Allan Hanson

***Anthropology and the Greeks**
S. C. Humphreys

**Available in paperback*

The demise of a rural economy

From subsistence to capitalism in a Latin American village

Stephen Gudeman

Department of Anthropology
University of Minnesota

Routledge
London

First published in 1978
Reprinted in 1981 and 1988
by Routledge
11 New Fetter Lane
London EC4P 4EE
Set in 10 on 12 point Compugraphic English Times
and printed in Great Britain
by T.J. Press (Padstow) Ltd., Padstow, Cornwall

British Library Cataloguing in Publication Data

Gudeman, Stephen

The demise of a rural economy.—(International library of anthropology).
1. Latin America—Economic conditions—1945-
2. Latin America—Rural conditions
I. Title II. Series
330.9'8'003 HC125

ISBN 0 415 01260 0 Pbk

Contents

Illustrations

Map

Tables

Figures

1 Anthropological economics and a small village

I have a story to relate about a rural economy in Latin America. At its simplest the plot concerns the way in which the people of a small Panamanian village began to leave aside planting rice and maize in order to accommodate a new crop, sugar cane. But this crop change entailed more than a reorganization of agricultural practices: the introduction of sugar cane represented a transition from subsistence farming to cash cropping, and this agricultural reorientation itself presaged a total transformation of the people's economy. Thus, more broadly, my narrative is about a rural economy caught in the moment of transition between two competitive forms of organization, production for use and production for exchange. Further, not only were these two patterns of production historically related, but the change from the one to the other was a consequence of forces emanating from without the immediate village. Perforce, an account of the local economy must stretch in time and space beyond the geographic borders of its ostensible subject. My story, then, revolves about a subsistence economy: to inquire into its essence is also to ask how it was generated by and maintained on the margins of a different economy, eventually to be absorbed by that larger system.

The concepts of economic anthropology hardly provide the means for relating this brief tale. The debate between the formalists and the substantivists, that acrimonious discussion frequently unrelated to ethnographic facts, has so numbed the entire field it scarcely seems that an anthropological economics is possible. The substantivists may have over-emphasized the domain of transactions, but the byway of the formalists is more treacherous, for they have abstracted from the entire history of economic thought one period, which lies roughly between Marx and Keynes, and raised it to the level of universality.

From one viewpoint both approaches have obscured the essential. Hidden by the smoke of the battle is the fundamental fact that both sides agree on the starting place for analysis: exchange relations. Ultimately, this leads both to encounter difficulties in conceptualizing the economy in relation to society, and aligns both with the same broad tradition in economics.

Since the apparatus of the formalists has been taken with some emendations from neoclassical economics, since their key text appears to be a short essay by Lord Robbins (1932), their interest inevitably is focused upon decision making and the 'rationality' of human beings as they meet to exchange. But by starting with the assumption that everywhere means are finite while ends unlimited, the formalists have achieved the dual 'victory' of universalizing economics to the point of absorbing all other forms of human activity while denying the anthropological premise. Humans do make decisions - Malinowski long ago suggested that rules and actions do not always coincide - but when the anthropologist becomes so infected with the study of decision making (Ortiz, 1973), when the economy, *pace* Adam Smith, is reduced to an aggregate of individual decisions, the anthropological perspective has been lost. Taking individuals, or elements, rather than groups, or systems, as the place of departure is to mistake result for cause, a fact long ago argued by Durkheim. Seemingly, the formalists would deny that an economy is a specific, determinant system, a system which is reducible neither to individual volition nor to a reflection of the formalists' own real-life experience.

Paradoxically, the same broad critique holds for the substantivists, for their principal theoretical propositions - some might say their only - concern exchange. For Polanyi (1968: 148-9) the way into an economy was to locate first its primary 'form of integration', which had to be reciprocity, redistribution or exchange (meaning for Polanyi market relations). Indeed, one is tempted to go further and suggest that if one element of a 'true' substantivism consists in the appreciation and use of native categories, then the substantivists have betrayed their own origins and cause by emphasizing modes of integration. Anxious as they are to display the social origins of the economy, the substantivists, because they commence with exchange, are forced to reinsert the economy in the society after first extracting it.

All this is not to say that, aside from the great debate, a generation of anthropologists has not raised important queries, but when we are told that the economy is 'embedded' in the society, we need to be offered something more than the broad strokes of a Polanyi, the vignettes of a 'mediator' such as Firth (1963: 147-52), or the micro-analyses of a formalist, such as Nash (1961). When it is asserted that a rural economy is connected to the national system of which it is a part, we must not be limited to looking only at the physical exchanges of the marketplace or that more abstract entity, the market (Smith, 1975). When we talk about subsistence we should cease using the term only to designate standards of living or a form of agriculture (Richards, Sturrock and Fortt, 1973) and realize that production for consumption is radically different from production for exchange. And the whole problem of surplus - what it is, and whether it exists in fact or theory - must again be confronted. We need, in short, to return both to ethnography and to the basics: 'Economics is a study conducted by human beings about human conditions. Mankind supporting life by labour is its central concern' (Robinson and Eatwell, 1973: 61).

My story is offered in this spirit of rejuvenation, of trying to build again an anthropological economics. Such an anthropological economics, as opposed to economic anthropology, should, as many have emphasized, be couched within the traditions of economics yet not forsake anthropology's unique capacity to expand the context of those same traditions. But upon what tradition in economics, a science forged originally in market economies, can the anthropologist draw?

We may begin with the observation that a central problem for any society revolves about the question of distribution: given a product, or products, who is to receive which goods and in what proportions? Broadly viewed, within economics there exist two contrasting approaches to the determination of distribution, two theories concerning the way in which goods are valued, or priced, and apportioned. One theory, sometimes known as neoclassical or marginalist thought, approaches the problem of price determination and distribution through exchange. Prices are set in the market, where ideally a multitude of buyers and sellers meet. In the neoclassical view income distribution - how much the labourer earns, what the rate of interest and return on capital are - is settled in the market. This is the familiar world of supply and demand curves, of the matching of marginal or last increments of inputs and outputs: wages are determined by the marginal product of labour, rent by the marginal product of land, and the return on investment by the marginal product of capital.

A second tradition, primarily associated with Marx but which actually derives from Ricardo and counts among its followers the 'neo-Ricardians', approaches price determination and distribution principally through the initial situation of production. Demand, consumer desires, here plays a role too, but in a less direct fashion than in the neoclassical version.

According to the Ricardian tradition, transactions or exchanges are based upon an initial distribution of resources, goods and labour. The command over resources - hence distribution - begins in the production process itself. For Ricardo 'the principal problem in Political Economy' was indeed to determine the laws which regulate the distribution of the earth's produce as amongst rent, profit and wages; but he began his treatise with the assumption that (Ricardo, 1951: 5):

> the produce of the earth . . . is divided among three classes of the community; namely, the proprietor of the land, the owner of the stock or capital necessary for its cultivation, and the labourers by whose industry it is cultivated.

The division of society into 'classes', the socio-economic backdrop, was for Ricardo an initial, given fact. Thus, in the Ricardian tradition the embedding of the economy in society occurs not at the level of exchange, the level at which both the formalists and substantivists commence and the level which Marx called the 'sphere of circulation', but is a condition of production itself (Dobb, 1973: 169).

Since pricing mechanisms are contingent upon the use of money and of markets, it might appear to the anthropologist that this disagreement among economists has little relevance to those societies lacking full-fledged markets, fully developed monetary systems and the like. In fact, the reverse is true, and this takes us directly to the theoretical importance of a subsistence economy.

In a subsistence system there are no marketplaces or markets. The producer works in order to have goods he may consume. Crudely put, worker and 'capitalist' are one. Apparently there is no division between return to labour and to capital. Yet, there still remains the analytical problem of determining what sets the consumption level of the labourer. Further, I shall maintain that even in a subsistence system, the worker produces a surplus, although he controls, more directly than in capitalism, the disposition of this surplus. The crucial question, then, is what determines the distribution between the subsistence level and the surplus in an economy lacking markets, in an economy where market principles cannot apply?

In the course of my description I shall suggest that the subsistence level, surplus and the production technique stand in a mutually dependent relation. Further, the distribution of output between surplus (roughly profit) and subsistence (roughly wage) is not made *ex post facto* in some supposed marketplace of the worker's 'rational' head; rather, the determination of distribution is inserted into production itself. Following the Ricardian tradition, I shall hold that distribution, the division between subsistence and surplus, is essentially a social and not an economic fact. To the outsider, the villagers' subsistence level appears to be an absolute economic standard, because conditions of productivity in the countryside change little, but essentially the division between subsistence and surplus is set by social conditions and expectations. For example, the time a man spends in his fields (which is supported by his subsistence) versus the time he spends paying homage to the saints (supported by his surplus) is part of the cultural fabric of the society and is not dictated by economic facts alone. To cite a more familiar illustration: by pressing for a larger share of the product, by asking for more pay or vacations or fringe benefits, unions not only are attempting to raise the real wages of their members but also are trying to change the conception of what a worker is in relation to a manager or capitalist. As Dobb (1973: 152) has remarked of Marx's theory: the result 'was to make income-distribution an historically-relative product of a given set of historical or institutional conditions'.

If, then, distribution is treated as an initial datum and not as an outcome of the market or reciprocity or redistribution, the emphasis of the analysis must shift from the sphere of exchange to production, and beyond that to the socio-cultural conditions in which the production process itself takes place. It follows that the anthropologist has a unique contribution to offer to the unravelling of the seemingly 'economic' problem of distribution.

Pride of place also must be given to labour as the ultimate factor of production, especially in the economy I am about to describe. But the Marxian concept of 'labour value' is notoriously difficult to use (Robinson, 1962). Certainly if emphasis is given to the role of labour within the production process, it is necessary to be wary of one's assumptions.

When employing labour as a cross-cultural concept, the anthropologist encounters that most difficult of transitions: the movement from appearance to 'underlying reality'. If the distinction between labour power (the individual poised to work) and labour (the actual process of working) is fundamental to Marx's analysis, and if this distinction is found only where labour itself is a commodity (as in capitalism), then how valuable is the Marxian apparatus in the analysis of non-capitalist societies? Are the ideas valid where labour is neither a culturally recognized category nor institutionally differentiated in the relationship system? In such cases does the labour process still provide the underlying reality of production, regardless of visible form?

We may phrase the problem slightly differently: to what extent do the Marxian concepts pertain to nature - the universal substratum - or to culture - the attribution of meaning by humans to existent objects and actions? The 'mature' Marx assumed that 'the property . . . which labour-power in action, living labour, possesses of preserving value, at the same time that it adds it, is a gift of Nature' (1967a: 206). But is this gift from Nature to man universally - naturally - true, and if not, then what can provide the foundation for a cross-cultural economics?

Shifting attention from the labour side of the expression 'labour value' to the second term, similar reservations hold. Labour value as analytical device must be disentangled from labour value as ideological banner. The distributional problem - who gets what and in what proportions - surely has to be solved in all societies, but in the first instance labour value ought to be conceived solely as the amount of work needed to yield an object, not as some 'mystical' relation between the worker and his product from which he is sometimes separated. Labour value is not an ethical notion which can provide the foundation for defining in the abstract a 'good' system. There exists no 'naturally' just method for distributing the rewards of work; all are based on moral, indeed cultural, presuppositions. After all, even in the completely planned economy, the labourer does not receive the total output of his efforts. Some portion of the product must be 'taken' to provide for accumulation.

Labour performed also ought not to be understood as constituting the origin of value in the sense of defining what the 'good' objects are. In any society not all human activity creates value, and that physical action which creates value in one society may not in others. Value in this respect also is ultimately a cultural conception. Exactly why cars, paintings or wines are prized in one society, armshells and necklaces in another, and brass rods in yet a third cannot be answered by calling upon some simple theory of labour value. To be sure, valued objects are produced by labouring, even if this labour amounts to no more than uncovering a precious stone, but the initial fact of 'value defined' precedes, shapes and moulds the labouring process.

Then, there exists a variety of measurement problems. Given a set of social and historical conditions, given the institutionalization or definition of value within a society, it is an open - not a settled - question whether the relative values of two produced commodities are in proportion to the comparative amounts of labour 'embodied' in them. Whether two objects should exchange in relation to the amount of labour they 'contain' is an ethical question; whether they do so exchange is an analytical problem.

But even if it were asserted by the members of a society that their produced objects ought to exchange at a rate relative to the amounts of labour they contain, there would remain the further question of how to measure this labour. How is labour performed by different human beings to be equated, and what unit of measurement can be utilized except time duration? Yet, time itself is conceived differently in the world's cultures.

None the less, given these limitations, the role of labour within the production process remains worthy of emphasis, for technically a product may always be evaluated in terms of its labour inputs. The product incorporates direct labour given over in the production process plus its raw materials and a portion of the equipment 'used up'. Both raw materials and equipment, however, are themselves products of prior expenditures of labour and tools. Since equipment represents 'frozen' labour, the value of productive instruments ultimately may be reduced to 'dated labour' (Sraffa, 1960) - labour expended in the distant past. But here we may usefully distinguish between two meanings of labour as a measure of value (Meek, 1973: 51). From an external and technical standpoint we can (theoretically) measure in time units the amount of labour which is embodied in a product. Whether internally, to a people, labour is the measure of value is a question of a different ontological order. In this second sense one is asserting something more than that labour is a yardstick or thermometer: labour is *the* measure of value because it is *the* source and substance of value. This shift in meaning is equivalent to a movement from the domain of the technical to the cultural.

Yet, it is precisely this transition from the one meaning of labour value to the other which should draw anthropological attention. Labour is both thought and action, and situated between nature and culture. As the activity of sustaining oneself, of interacting with nature, labouring is an inherent human condition; to persist, mankind must provision itself, both produce and reproduce. Culturally, however, there is great variation in the conceptions specifying which human acts are defined as labour, where and when labour is deployed (its temporal and spatial framework), how it is measured, the objects to which labour is applied, and how the results of labour performed are to be evaluated. One merit of a comparative economics is that the anthropologist can show both what labour means for a people and how distributional mechanisms are related to labour inputs. Exactly why the concept of labour value has had both an equivocal and a slight reception in anthropology is perhaps for an historian to answer.[1]

Given the conditions of the Panamanian countryside, it is pertinent to focus on labour as a productive input, for several reasons. Most of the essential goods produced in the subsistence system are created directly by labour, labour almost unassisted by equipment. By using labour as a measurement of the product one also can assess the technical process of production and the productivity of the system. And when shorn of the polemics, the concepts of labour value and worker's subsistence provide a new understanding of surplus. Even more, the concept of return to labour provides a welcome bridge for comparing two different economic systems, subsistence and capitalism.

My story, then, concerns the village of Los Boquerones, as it existed in the years 1966–7. Los Boquerones lies in Veraguas Province, in the Panamanian heartland. I originally chose the community for study precisely because it seemed to be undergoing a change, though I scarcely understood what this change was at the time. The village contains 350 people and 91 households, and is rather typical of the 'peasant' population in Panama.[2] The people are Spanish-speaking *mestizos* whose culture resembles that found in many parts of Hispano-America.

To describe the economy of Los Boquerones in 1966–7, however, is a rather complicated venture. At each stage of my narrative I have had to hold certain dimensions constant while describing others. Each section of the story represents only a partial unfolding of the entire picture, and sometimes the segments are viewed in change while other times they are treated almost fictitiously, as timeless entities. Only one of many descriptive problems I have faced is that the economy of Los Boquerones now is not one but two inter-digitated systems. Both are there at the same time, but to understand them, I have had to separate the two and describe each as a complete system. If, at the end, the reader concludes that 'the economy' of Los Boquerones is a contradiction, that it could not persist as it was in 1966–7, my mission will have been successful.

Throughout I have tried to combine both macro- and micro-levels of analysis, in a double sense: macro-theory linked to micro-details, and the macro-economy of Panama connected to the micro-economy of Los Boquerones. I begin with an historical sketch, because the nature and direction of the current Los Boquerones economy is an embodiment of that material history. The subsistence economy of Los Boquerones is an isolate, a system, but it was generated by and is defined in relation to a broader system of which it is a part but which itself has decidedly different characteristics. The principal form of control by the 'outside' economy comes through its ultimate governance over the means of production in Los Boquerones. Because this is an historical fact, history provides a key to grasping the relation between village and metropolitan area.

In Chapters 3 and 4 I turn to the subsistence economy as a functioning entity; the first of these two chapters concerns what might be called the conditions of production while the second focuses on the actual processes of production. By conditions of production I refer to the social organization which underlies production. The production process refers to that complex formed by environmental possibilities, technology and actual methods of producing objects. The two facets, of course, are only analytical frameworks for describing that which is a unity. Since, aside from the pioneering efforts of Sahlins (1972), little anthropological attention has been devoted to subsistence as a distinctive form, one of my purposes in these chapters is to display the special nature of a subsistence economy and its relation to the local and national social orders. Here also I turn to the old issue concerning the 'existence' of a surplus. Distribution, the technical relation between the surplus and subsistence portions of the product, I shall show, is ultimately a result of social causes. In different language, the 'embedding of the economy in society' occurs 'first' in production.

Following this discussion of subsistence production I 'surface' to the level of circulation, examining that most important of exchangeable items, human labour. The exchange of labour within the subsistence economy is, in a sense, 'secondary' to the fundamental task of securing one's livelihood through production for the self.

But the haunting theme of change, change which results from the incursion of an outside system, appears time and again through all these chapters. Thus, in Chapter 6 I turn to the most explicit and current embodiment of that change, sugar cane. By 1966–7 the 'forces of modernization' were most directly represented by two nearby, privately owned sugar-cane mills. Ostensibly, changes were occurring in the community due to increased sales and purchases on the national market; in fact, the crucial element in this change was the mills' enlarged control over local production through their command of a financial fund, or a new distribution of productive resources between countryside and city. But this view of the transformation from subsistence to capitalism is entirely consistent with my earlier analysis, for in both the economy is viewed from the perspective of production. Ultimately, the switch to sugar cane was an extension of the same historical factors which produced the subsistence system.

The epilogue brings the story up-to-date. Transformations and continuities have continued in Los Boquerones since the mid-1960s, but now they are occurring under the impact of political power rather than corporate influence.

2 An economy evolves

Although it was generated by capitalism and proto-capitalism, the subsistence economy of Los Boquerones is not itself capitalistic. Producer, means of production and product are not separated, free labour and a labour market are not found, and surpluses are not cumulated and reinvested. But unlike the Asiatic, Germanic or feudal modes which are said to precede capitalism, the peasants' economy is concurrent with it. The problem, then, is not that of determining whether this rural structure is 'feudal' or 'capitalistic' (Frank, 1967, 1969; Laclau, 1971) but of comprehending it in relation to the larger economy of which it is a part.

Existing on the margins of the capitalistic economy, the rural system contains within it the 'pre-conditions' for conversion to that 'more advanced' form of production. The crucial link between the two comes not through the market, by means of the exchange of goods or labour, but through control over real property. The rural economy evolved in a context where the rights to the natural wealth of the area were held by small-scale capitalists who lived in the large towns of the countryside or cities of the Isthmus area. So long as these owners had little interest in the resources of the countryside, some of the productive means remained free and could be utilized by the local populace. Hence, an independent, subsistence system was able to grow up and then flourish on the edges of capitalism. Little surplus was extracted from this rural system, yet never was it permitted to flower into an opposing or competitive economy. Thus, as soon as the resources of the countryside became valuable, the dominant property owners had only to tighten their grip on the means of production, and the subsistence economy, as a functioning system, began to be squeezed out, even extinguished.

The current Los Boquerones economy, therefore, is an historical precipitate, to be understood not only as it now is but also in terms of the conditions in which it was created; furthermore, the present system is changing dramatically, not 'in itself' but in response to the same historical forces which produced it. Neither a holdover from pre-Hispanic times nor an economy independent of the Isthmus, the subsistence system developed, paradoxically, out of mercantile capitalism. In the early 1500s the Panamanian interior was subjected to a rapacious search for treasure; later its gold was mined by the Spanish; finally it became only an appendage to the more important transit area. The economy of the interior, like a small puddle left drying in the sun, is unintelligible by itself: isolated, with no direction in which to flow, but a logical, inevitable result of the preceding storm.

The post-Conquest history of the interior may conveniently be divided into three phases, phases which successively narrow toward Veraguas Province and Los Boquerones. These epochs are not formal political periods, such as the Colonial, Colombian and Independence eras, although the shifting world sources of power certainly had their influence on the countryside. But neither are the phases strictly economic periods. Throughout these years the developing capitalistic system initiated changes and the rural area responded, but the shifts occurred in property rights, instruments of production and sources of value. The first phase of conquering and exploiting of raw materials and labour began with Columbus and lasted for over a hundred years. This set the stage for the fallow or classic period in which the subsistence system evolved on the margin of capitalism. Finally, beginning in this century irreversible impacts from the transit area began to have their effect upon and dissolve the well-established subsistence economy. Accelerating in the last twenty years, these forces have led to the current subsistence/cash crop system but eventually they will engulf the countryside in capitalism.

Great expectations

Lasting until the mid-1600s, the first phase saw the discovery and conquest of the interior by the Spanish. This epoch marked the only time when, for a brief period, the countryside was the axis of, rather than an appendage to, the Panamanian economy.

The first European to land on the Atlantic side of Panama was Rodrigo de Bastidas in 1502, a member of whose company was

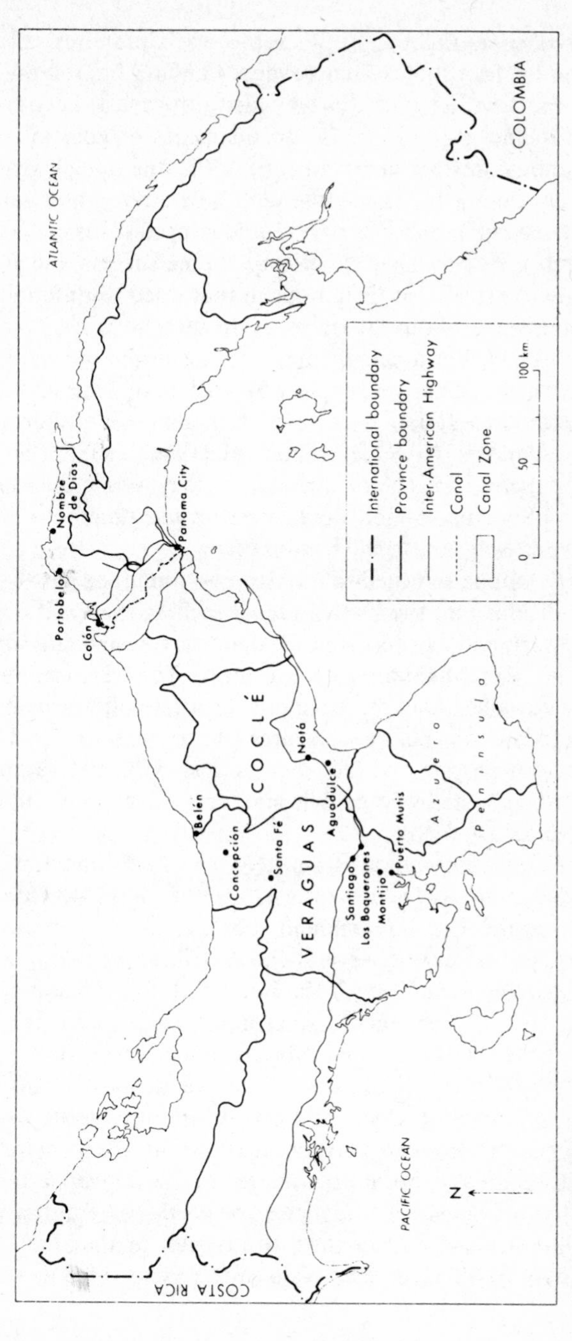

The Republic of Panama

Vasco Nuñez de Balboa. In this same year Columbus sailed from Cádiz on his fourth and final voyage. Landing first on the eastern part of Panama, he made his way along the coast, hearing - from various Indian groups - of fabled quantities of gold in the more mountainous, western areas. In early 1503, due to bad weather, he landed on the north coast at the mouth of a river, and named the site Belén (Map 1). Soon a party, guided by Indians, set ashore in the search for gold. They experienced some success, but it was an unfortunate success for Columbus in that it led to unfounded fantasies on his part about the riches of the land he called Veragua (to see water).[1] Within a short time, fighting broke out between the colonizers and the natives, and the Spanish fled. Thus, Columbus's first attempt to found a colony on Panamanian shores was a complete failure. As Sauer remarked (1966: 136): 'Years later, when the claims of his heirs were settled, they were given the title of Dukes of Veragua, a land their ancestor was unable to hold and which they never saw or had benefit of.'

In 1510 Balboa returned to the Isthmus and in the next two years led expeditions into the Darién area, reaching the Pacific in 1513. Unlike nearly all his successors, Balboa established amicable relations with the inhabitants of the land. To the west, however, Veragua remained unexplored and its borders undefined. A second attempt to land and colonize the area was made in 1510 by Diego de Nicuesa, but despite the fact that he had with him sailors from Columbus's ill-fated voyage, Nicuesa lost both his way and most of his expeditionary force.

The first governor of Castilla del Oro (the Panama area), one Pedro Arias de Avila (Pedrarias), arrived in 1514. One of the sorriest figures in Panamanian history, his accomplishments included the massacre of several Indian tribes, the taking of slaves and gold, the execution of Balboa, and the founding of old Panama City on the Pacific side of the Isthmus. In the second decade of the 1500s the Spanish began a new westward movement, but this time by land from the Isthmus area, mostly into what is now the province of Coclé and the Azuero Peninsula. Within a short time thirteen *entradas* or attempts at the penetration of Veragua were made, but no permanent footholds were established, due partly to Spanish incompetence and partly to fierce native resistance. The first permanent town to the west of Panama City was Natá, founded in 1522. The Natá area was to serve as the entry point to the interior for years to come and as a source of some food supplies for Panama City.

In 1526 the successor to Pedrarias began a survey of the Isthmus, a survey which eventually led to the first recommendation that a canal be constructed between the oceans. Although this vision was not to be realized for nearly 400 years, in 1520 an inter-oceanic trail was begun between the town of Nombre de Dios on the Atlantic and Panama City on the Pacific.

When Pizarro led his expedition to Peru in 1532 and shortly after began to remit gold and silver to Spain, the importance of the Panama route - in preference to the Cape Horn passage - was firmly established. Thus began what might be called the economic bifurcation of the country. The Isthmus itself became a trade route intimately linked to the world economy; the attraction of or lack of interest in the interior was for different reasons.

Natá, to the west, was founded on the *encomienda* system. Spaniards were granted rights to specified areas of land and to the labour of a certain number of Indians within that area. But this method for extracting the labour of others never reached in Panama the scale that it did elsewhere in the Spanish world, and in the Natá area *encomiendas* were officially ended in 1558. In the following year, however, a short new period began in the interior with the founding of Concepción, a mining town on a river in northern Veragua. Having looted the gold which had already been produced by the Indians, the Spanish turned to exploitation of the raw material, using imported slaves to do the work. Concepción soon numbered over 2,000 inhabitants, and in 30 years some 2,000,000 '*pesos*' of the precious metal were mined (Castillero Calvo, 1967: 56).

Not being located in a fertile area, Concepción had to be supplied with agricultural produce and meat from elsewhere. Its riches gave rise to farming in the interior. Thus, when the *encomienda* ended in Natá, there was a dispersion of small-scale farmers from the town into the savanna areas of the Azuero Peninsula and what was then Veragua. This local economy was based on the production of cattle and maize, a duality which still persists in the interior. But individual farms were small and only loosely knit together, and the inhabitants traded little with the Isthmus area.

By 1589 the mines of Concepción were depleted. For a short time thereafter the centre of mining activity shifted to Santa Fé, high in the mountains of northern Veragua, but soon many of the miners, along with their slaves, left for the Zaragoza district of Colombia (West, 1952). With the collapse of the mining and its market, the economy of the interior underwent an involution. The only other potential market - the transit area - was not only self-sufficient but able to supply the visiting fleets with foodstuffs. It had no desire for competition from the interior; in fact, in the 1590s an over-supply of cattle in the Isthmus led to a drop in meat prices. Since the abattoirs were controlled by the cattle owners of the cities, to boost prices they squeezed out their rural counterparts. This economic collapse of the interior led to a further population dispersion on to the Pacific plains and savannas in the late 1500s. As Castillero Calvo (1967: 79-80), upon whose work I have drawn, commented:

> The rural expansion, provoked by the mining crisis of 1589 and consolidated during the 17th century . . . would establish one of the characteristic traits of Panamanian society to our days . . . the dispersion of the *campesinos* throughout the [interior] . . . in small families, at times in small primitive hamlets and practically marginal to the monetary system.

To the west of Natá the *encomienda* system was re-established along with slavery, and for a short time the interior did supply the transit area with dried beef. In contrast to the well-trodden overland Isthmus route, however, transportation between the interior and Panama City was usually by sea; cattle products were sent by boat from one of several Pacific ports rather than on hoof (Castillero Calvo, 1967: 122-3).

But even this minimal production above local need was short-lived. Individual *encomiendas* were never very large, and beginning in the 1620s various Crown edicts began to abolish the practice (when a proprietor died his *encomienda* reverted to the Crown rather than to his heirs). In 1620 some *encomiendas* had 40 Indians, but by 1630 the largest contained 30, with most having less than 10, and by 1650 the system had disappeared (Castillero Calvo, 1967: 121). Although Panama City in this period actually was in need of foodstuffs, rural agricultural production became stagnant. The Indians remaining in the interior produced only for themselves, and Indian-Spanish commerce was reduced to the exchange of handicrafts for European trinkets.

Conditions in the interior appear to have been much like those of today: production was for the home, labour service of others was not utilized, houses were simple, a few cattle were to be found, and maize was the mainstay of the economy.[2]

The current capital of Veraguas, Santiago, was founded about 1636. Located midway between the mountain town of Santa Fé and the seaport of Montijo, it occupies a geographically central position. By 1691 it contained 1,000 persons and was the centre of the interior, a position it preserved throughout the colonial period.

Who actually held rights to the land of Veragua is not clear, Columbus did not succeed in occupying it. In 1508 Nicuesa was granted the western part of Tierra Firme, known as Veragua, which according to Sauer (1966: 170) ran from the Gulf of Urabá in Colombia to Honduras. But, as noted, Nicuesa's attempt at colonization also failed. In 1513 King Ferdinand declared that Veragua belonged to Diego Columbus, son of the mariner. He, too, never occupied the area, and after new negotiations Luis Columbus, grandson of the explorer, received among other emoluments the title Duke of Veragua and in perpetual fief the Duchy of Veragua. In 1556, however, the Crown forced Luis to give up his rights. The Concepción mines were worked under a syndicate organized in Natá by Francisco Vasquez de Coronado and chartered by the King in 1557 (Lothrop, 1950: 10). But the actual land in Veragua in the late 1600s remained in the hands of a few, though given the paucity of population and slight exploitation of resources it could hardly have been exceedingly valuable (Castillero Calvo, 1970: 73).

Political power, often obtained by purchasing public offices from the Crown, buttressed wealth, and certain families were able to retain their landholdings through time. In addition, a differentiation began to be drawn between, on the one hand, a wealthier white populace, who held political offices, and on the other, poorer *mestizos*. How developed this socio-economic differentiation became is not clear, but at least the seeds were cast for a rural-urban, agriculturalist-cattleowner, landless-landholder, *mestizo*-white distinction which, to some degree, is still in existence today.

Thus, by the middle of the 1600s, the first phase of European settlement in the interior of Panama drew to a close. It had opened with the inflated hopes of Columbus. The first sea expeditions to Veragua failed, but the overland *entradas* succeeded in obtaining treasure and in harnessing Indian labour. Later, the Spanish turned to mining gold, but when the mines were exhausted the economy relapsed and Veragua, isolated from the Isthmus, turned in upon itself. The settlers moved to a better ecological niche, and a fallow (or perhaps classic) phase of existence began in which the essential structure of the interior took form. The economy shifted from one of exploitation to production for use: labour was no longer extracted from others, either in the form of slavery or *encomienda*, while households produced for themselves. Except for the addition of cattle and rice, the settlers used primarily the crops and agricultural techniques which the Indians had employed for centuries prior to their arrival. Any further changes in the rural economy would have to emanate from the metropolitan areas of the transit zone, but this area too was soon to experience an economic decline.

Relapse and resurgence of the Isthmus

The fortunes of the Isthmus had varied with world conditions. Originally, Nombre de Dios, the Atlantic terminus, was one of three ports authorized by the Crown to trade with Spain and to receive the annual fleet. Shipments from western South America to Panama City were sent in time to be transported to the Atlantic port and thence to the fleet for its return journey. Panama itself contributed little to the load, and hence its prosperity fluctuated with the trade volume. Spain was not able to retain its trading monopoly for long and in the late 1500s Francis Drake made innumerable assaults on the north coast. None the less, a peak of prosperity was reached in the early 1600s. Portobelo, which had become the successor port to Nombre de Dios, was the site of large fairs where European merchandise for the entire west coast of South America was displayed. On the Pacific side old Panama City became one of the more affluent settlements of Latin America.

By the mid-1600s the volume of precious metals fell. In 1671 the buccaneer, Henry Morgan, sacked old Panama City, although two years later a new city was re-established six miles away. By the late 1600s Spain's power had declined, and though the Portobelo fairs grew smaller, goods from other countries still travelled the overland route between the oceans. Politically, Panama lost its independent status to Peru in 1718. By the Treaty of Utrecht (1713), Great Britain received the right to send one ship a year to Portobelo, but continued smuggling undermined the official trade. In 1739 Portobelo was taken by the British and its forts were destroyed. Direct trade between the colonies and Spain took place and the Portobelo fairs were never again held. Thus, in the 1700s the Isthmus itself experienced an economic decline, since it no longer could reap the profits of colonial trade and had no other developed economic base. It produced nothing for export and provided little stimulus for the interior.

On 28 November, 1821, following Bolívar's victories to liberate New Granada, Panama declared its independence from Spain. The new state formed a political union with Colombia, and was itself divided into two provinces - Panama and Veraguas. Along with Ecuador, the entire area comprised Gran Colombia.

Left in charge of the Isthmus before liberation was one José de Fábrega, a native Panamanian and governor of Veraguas. General Fábrega, however, helped lead the revolution. Fábrega family ownership of land in Veraguas dates at least from this time. It is on a part of their land that Los Boquerones is situated.

In 1826 Panama was the site of a congress called by Bolívar, who was considering building a canal, but interests of the USA in such a venture also began to surface. The Panama Railroad Company was organized by New York financiers in 1847, and the first train ran between the oceans in 1855. In the same period the California gold rush did much to resurrect the Isthmus economy.

The union with Colombia, a country to which Panama still has no easy land access, soon turned sour. Between 1830 and 1841 alone Panama made 3 separate attempts at secession. Throughout the rest of the century internal problems and struggles with Colombia continued, and the USA began to play an increasingly important role in domestic politics. Following a revolt in the new Atlantic terminal city of Colón, for example, US troops in 1885 briefly occupied both Colón and Panama City, and guarded the railroad.

Ferdinand de Lesseps in the late 1870s formed a company to construct a sea-level canal, but though excavations began - work which was to be useful to the later canal - by 1889 his company went bankrupt. In spite of the French failure, the attempt brought new economic impulse to the Isthmus, as well as a labour force of Jamaicans.

The War of a Thousand Days (1899-1902) in Colombia, which exhausted the country, brought the Isthmus to a final stage before Independence. In early 1903 the USA signed a treaty with Colombia which would have allowed it to build and control a canal, but later in the year the treaty was rejected by the Colombian Senate, which wanted larger monetary payments and greater control. With the support of US naval forces and the offices of a French citizen, Philippe Bunau-Varilla, who held stock in the French canal company, Panama declared its independence from Colombia on 3 November, 1903. On 14 November the USA gave legal recognition to Panama. Bunau-Varilla then negotiated and signed a new treaty with the USA before a Panamanian delegation, which had been sent to the USA, arrived. Work on the Canal was begun in 1907, and it opened in 1914. The stage, thus, was set for a third phase in the interior and a new penetration.

New forces in the interior

The opening of the Canal marked the beginning of a new period in Panama. The Canal Zone itself has assumed increasing economic importance for the terminal cities and nearby areas but has had an impact on other sections of the country as well.

In the early years of this century the USA frequently intervened militarily in Panamanian affairs, a right it held until 1936 by the original treaty.

Internal political disturbances in this century also have been numerous, although political power until very recently remained exclusively in the hands of a metropolitan elite. This elite, however, often retained an economic base in the countryside. Two members of the Chiari family, Rodolfo (in 1924) and his son, Roberto (in 1960), have served as president of the country. This same family owns one of the two large sugar-cane mills to which the peasants of Los Boquerones sold their raw cane in the 1960s. The family of the vice-president, elected in 1964, owns the other mill.

As the tentacles of industrial capitalism have reached into the countryside, the local economy has begun to change. Although the changes themselves are piecemeal and only partially understood by the rural dwellers, they arise from the clash of two divergent economies and have an overall pattern and inevitable conclusion. These changes may be divided into two types, of differing importance. The more significant are those which may be classed as 'factor push': the depletion of natural wealth in the countryside and increasing control exercised over rural property and production by urbanites. But the groundwork for these changes has been laid by the second set of elements, 'market pull', for the rural dwellers, defining themselves as Panamanians of low social and economic position, are also receptive to those elements and symbols of modern life which characterize the metropolitans. As their economic base is cut from beneath them, the peasants seem prepared to follow the goals of the larger society, with some misgivings.

Population and transportation

Setting the stage for these changes has been a marked population increase in the countryside, although it may be that the current population is just beginning to match that which occupied the land prior to the arrival of the Spanish. Estimates of the pre-Conquest population have varied widely; Bennett, reviewing these on the basis of ecological changes, suggests that in the late 1400s there may have been as many as 1,000,000 natives on the land (1968: 37). This population was greatly reduced by warfare, disease and the privations of the *encomienda* system. By the late 1600s, according to Castillero Calvo (1970: 83), Veragua, which covered some 30,000 square kilometres, contained less than 3,000 persons, including *criollos*, *mestizos*, and Indians. Although by the late eighteenth century this number had been augmented sevenfold, in the mid-1800s Veraguas numbered only 33,000 inhabitants, still a low density. The real increase in population did not occur until after the turn of this century, undoubtedly the effect of the eradication and control of various diseases, the decline of infant mortality, and the building of hospitals in the countryside - all direct consequences of the US presence. From a population of 59,614 in 1911, Veraguas increased to 151,849 in 1970 (Bennett, 1968: 57; *Censos Nacionales de 1970*, n.d.). Of course, this population is not evenly spread across the land surface, and with increasing amounts of land being sequestered by the cattle owners, this minor population explosion has begun to have telling effects on the economy of the peasantry. In the case of Los Boquerones the problem is growing acute. At the turn of the century the land was scarcely occupied;

then, the community began to grow as people streamed in, mostly from the north, in the search for usable land. By 1967 the population of 350 people represented a density of more than 13.5 persons per square kilometre. Such a population, given the ecological conditions of the area, can be maintained by the traditional system of agriculture, provided that other conditions remain constant, which they have not.

Perhaps the most graphic symbol of the increased interest of the Isthmus in the interior, and a crucial infra-structural change in its own right, has been the development of the road network. Certainly, even prior to the Spanish, Panama was the roadway between the Americas. But until this century the actual land surface had been changed little. For example, one of the Fábrega descendants in the early 1900s drove oxcarts between Santiago and its nearest port village (Puerto Mutis), hauling goods which had been shipped by sea from Panama City. About 1913 local businessmen in the larger port town of Aguadulce (to the east) opened a dirt road between there and Santiago. But the trains of oxcarts often would get stuck in the mud along the way. On foot the trip to Panama City from Santiago took three to four days; very few peasants made the journey and the more popular, but little used mode, was to go by sea from Aguadulce. In the early 1920s an all-weather road was completed between Panama City and the interior, but it was not heavily travelled and the peasants still relied on foot or horse. The concrete Inter-American Highway was completed to Santiago only in late 1958. Soon plied by trucks and passenger vehicles, it had important and immediate effects. At last, the interior became accessible economically; its products - rice, cattle and sugar cane - could be sent elsewhere, and its people became a potential, albeit small, consumer market as well as a possible labour force. Following completion of the trunk route, feeder roads were constructed in the next decade. In short, the development of the road network in the late 1950s, financed ultimately by the US Government, and thought by planners to be a non-ideological benefit, made accessible the natural and human wealth of the interior in a fashion never before observed in this traditional passageway.

Land

The most valuable form of property in the countryside is the land itself, and control of this instrument of production has become increasingly important in the interior. The Los Boquerones land situation has its own particular history, but it is a history which is more than common in Veraguas. Typically, the land was owned by outsiders, living in Panama City or an interior town, who would use it to raise cattle. With the local peasants they would establish an almost symbiotic relationship: in return for rights to use a plot, the farmers would clear and help maintain the pasture of the owners. For a long period there probably was a convergence of interests between the two groups.

The original holdings of the Fábregas, which included Los Boquerones, were probably rather extensive. Though title was passed down in the family, some portions of the land were sold over the years. In the 1920s the family had perhaps 3,000–4,000 head of cattle in the Los Boquerones area plus a summer home, although their permanent residence was Panama City. They retained a foreman who oversaw the herd and employed a few labourers. Much of the actual pasture work, however, was done intermittently by the peasantry. When in need of forest land, to be used in the two-year swidden cycle, a farmer would request permission of the foreman to use a specified area. Upon receiving authorization, he would construct wooden crosses about the perimeter to mark his use rights *vis-à-vis* other peasants. Later, to keep cattle and hogs from damaging the crops, the farmer might erect temporary wooden fences. But these rights to use the land were impermanent. Only crops which could be harvested in a year or two might be planted, and eventually land farmed by one man might be worked by another. The Fábregas permitted the *campesinos* to build homes in the area, only if they were of the impermanent, stick and thatch, variety.

The fact that the Panamanian peasantry appears to be rootless, to move so often across the landscape – a fact frequently bemoaned in the city – is itself a direct function of this system of land tenure. The peasants throughout the interior were prohibited from becoming capitalist farmers or graziers since they could never make a permanent investment in (nor did they own) the principal means of production, the land.

In return for these rights of use the peasants would abandon their plots after two years, often seeding the land with cattle grasses as they left. The resulting savanna was then utilized for grazing. In addition, the agriculturalists paid a small rent each year, either $1.00 or a day's labour. (By the 1940s this rent was raised to three days or $5.00.) The size of plot made no difference; in the early years of this century, at least, a person could farm what he wished. The principal benefit to the landowning cattle owners, thus, was not a rent of money or food but labour service which was used to convert resources. This pattern of land use and payment helped maintain the peasants at a low level of production but did not in itself place much strain on their subsistence economy through the extraction of surpluses.

Thus, the people of Los Boquerones, and other parts of the countryside, were not participants in a feudal economy nor did their farming system resemble small craft production. Their subsistence economy was embedded within capitalism. Land was privately held, rent was paid for its use, bonds of personal dependency were not formed between owner and peasant, and a local farmer was free to shift his residence as desired. From within, no peasant had greater or more permanent rights to use particular plots of land than anyone else.

Whether the slash and burn agricultural system was ever in balance, such that enough forest was always being regenerated for use, is difficult to ascertain. Bennett (1968: 32) suggests that the burning of the forest cover has a long history in Panama, predating the Spanish by centuries. After the Conquest, with the dramatic population decline, much of the savanna reverted to secondary forest, and this condition prevailed directly up to the turn of this century (Bennett, 1968: 32, 55). In the last half century, however, the situation has altered, mostly as a result of the population expansion and the increased commercial importance of beef and agriculture in Panama. The inevitable result has been that the land has become more 'valuable'. Thus, throughout Veraguas over the last twenty years there have been many reports of squabbles - some major, some minor - between landowners and peasants. It is sometimes suggested that peasantries in general, the Panamanian rural folk in particular, are politically passive. But the one issue over which they may take up arms is the right to use the land, for when the peasants are deprived of the land, they have lost their principal productive resource.

In Los Boquerones the problem of land possession reached a peak in the 1950s, although the community still is experiencing the after effects of the crisis. When, in the early 1950s, the Fábrega owner died, the land was divided into five parcels for distribution to his heirs. Since by this time the Fábrega cattle herd had dwindled, rather than further splitting the land, the family decided to sell it. They began negotiations with several cattle owners, one of whom, before the deal was fully consummated, constructed wire fencing about his proposed plots. His actions provoked a semi-revolution in the community. The villagers tore down the fences, an act which in turn brought out the local police force and resulted in mass jailings. After an appeal to the president of the country, the offenders were released. Then, following nearly a decade of appeals and controversy, in 1963 the land reform agency, with the financial backing of the US Government, bought the land for eventual distribution. The total area of 2,632 hectares, purchased for $79,000, covers slightly more than Los Boquerones itself. According to government classification, 44 per cent is 'class II' land, 45 per cent 'class IV' and 10 per cent 'class VI'. By government assessment, however, the actual value of the land is $100 per hectare, rather than the $30 for which it was purchased.

By late 1967 the government plan for distributing the land was in readiness. It proposes to divide the area into a strip of house sites, bordering the highway, and into agricultural plots in the remainder of the area. House plots will sell for $100 per hectare, while agricultural plots will cost $30. The terms are to be 20 per cent down with up to twenty years to pay the balance. Existing rights of possession will be recognized to the extent possible. But farm land is to be divided first into ten hectare plots, and only when each family in Los Boquerones has been given the opportunity to secure land will further land be sold to initial purchasers or outsiders.

Liberal in form, the programme in fact has many drawbacks. Ten hectares are barely sufficient for operating the swidden cycle, not to speak of keeping several head of livestock. If sugar cane is planted, which it already has been, then the size is without question insufficient. Commercially, the small-sized plots are regressive in that mechanical techniques cannot be used profitably on such minifundia. Since in the traditional subsistence economy, the volume and velocity of money are small, the peasants themselves have quickly pointed out that few of them have the resources to make even a downpayment. Land purchase will force the *campesinos* to enter the capitalist economy, but plot sizes will not enable them to do so advantageously. The people will be liable for a debt for which they have no resources to pay.

Effectively the peasants, by a circuitous legal route, will do no more than buy the land from the previous owners. By law the agrarian reform agency cannot sell terrain to *campesinos* at a price lower than it paid. The land agency, therefore, has turned out to be not a force for land reform but only a marketplace intermediary. Worse, the government purchase was made in Panama City, and by this manner of transfer the final purchasers, the peasants, are going to have to pay a price over which they had no power to bargain.

Complicating the situation is a sub-legal, informal system of property rights utilized by the peasants themselves. Traditionally, the people retained and exchanged, by purchase and sale, objects related to the land but over which they had no legally recognized rights. The prices at which such possessions were exchanged did not correspond to their market evaluations as determined by the national system of supply and demand. For example, if a man planted a number of fruit trees around his house and then abandoned the site, he retained rights to the trees, although not the land on which they were planted. The fruit would be his, even if another person utilized the land surface for living or raising crops. He also retained the right to sell the trees. Similarly, houses or partially worked fields might be sold. The purchaser was buying, to use the latter case, not the land but the labour improvements and monetary costs added to it: cutting down the forest, fencing and fence materials, seed, the work of seeding, and so forth. In the people's terms, the value of an object was based on '*mano de obra*' - man power or labour. Locally, prices of items were determined by the amount of labour embodied in them.

In the case of the land, this labour theory of value came into a clash with the market system. Early in the 1940s, while the land was still owned by the Fábregas, the peasants began to utilize barbed wire instead of wood to encircle their temporary plots for protection from the cattle. The wire was cheaper and quicker to emplace. Later, when it became apparent that the Fábregas were leaving, the fencing took on a more important function. It became a marker of possession rights. Throughout the 1950s a local enclosure movement, agglomerating and encircling blocks of land, took place. Of course, fights arose over rights to areas, and such fights persisted regularly right up to 1967. The vacuum created by the withdrawal of the landowners gave rise to a situation in which different kinds of competitive claims to land were voiced. Thus, it was said that a person had rights to a land area depending upon how long he had worked in the general area, or upon whether he had seeded it, or upon whether he had fenced it. And to make the situation even more murky, in 1964 an outsider, claiming to be a legal surveyor, offered to survey personal land for $20 a plot. Those who could afford it accepted his offer. But no written records exist of the man's visit. The exercise turned out to be no more than a trick for extracting money and has no legal validity. None the less, among the people 'surveying' confers yet another type of claim to the land.

Currently, therefore, when land is exchanged among the peasants, the price paid is a combination of both its labour value and an amount added for its scarcity. For example, one man bought from another a small house site and house. The price of $25 was paid for the wire encircling the plot, the house materials and work put into it, plus a little bit extra for the seller's right of possession. The land, therefore, under conditions of growing scarcity, is being imputed with a value over and above the labour embodied in its improvements. This 'extra' amount now paid for the land might be seen as an inchoate Ricardian rent equalizing its marginal productivity.[3]

This new, scarcely defined, system of land rights is just beginning to have an effect on the economic structure of the village. One aim of the land reform agency is to allow people to continue to possess, so far as possible, the land they presently are using and improving. Since the people no longer value the land only on the basis of labour improvement but also scarcity, a new and slight economic differentiation among villagers is beginning to surface. The wealthier have been able to construct more fences and to purchase small plots from those in need of cash. When the land sale does take place, the effect will be to codify legally a mode of distribution thought to be established on the basis of use but which is not in fact. To an extent, some of the enterprising peasants purchasing land are practising a kind of arbitrage, translating goods from one sphere of value to another. Not surprisingly, two of the village families who left the community for Panama City, but retained their land in Los Boquerones, have established with other villagers a relationship exactly replicating that of the previous landowners. One family allows a villager to farm its land provided he plants sugar cane at the end of two years; the other requires that the users after two years seed cattle grass for livestock which they intend to graze.

Such a changing, fluid situation cannot easily be summarized as of a particular date. But the overall historical transformation is clear. With respect to the land, the peasants are in the process of moving from a system of value and price based on labour to a system of value and price based on supply and demand. Further, their market evaluation of the land, related however tenuously to potential profits, does not correspond with the government evaluation of the land.

Resource depletion

Added to the land problem, indeed a cause of the encirclement movement, is a diminution of the material wealth of the area. Occurring in Los Boquerones is the concentration, the impact, of trends which are taking place throughout the interior. The local combination is unique, but the changes themselves are far beyond the control of the *campesinos*. My own observations and data from informants conform closely with the macro-changes Bennett reported. His overall conclusion merits citing in its entirety (1968: 55):

> The period 1501 to 1903 was for its greater part marked by the ecological retreat of man and the re-establishment of forested or wooded conditions over much of the isthmus. The trend was retarded during the latter part of the period and then reversed, beginning with the modern historical period, when a resurgence of human population and new cultural and economic activities led to the mounting of a new ecological assault.

Exploitation of non-domesticated animals in Los Boquerones has practically ceased. Even middle-aged informants used to hunt wild pig, deer, iguana, and various birds. Hunting has become a sport, and the hunter must now travel to places near the Azuero Peninsula to shoot. Fishing of the streams and rivers is still practised but on a small scale.

Raw materials, too, have diminished rapidly in the past ten to fifteen years, and their shortage is creating problems. For example, firewood often has to be purchased, while large logs for house building are now a market commodity. Fine grass for thatching houses is nearly non-existent, while vines for binding are quite hard to find.

Domesticated animals - cattle and hogs - also have declined in number. By every report, small hogs used to be kept by nearly every household; sometimes a home had as many as ten to twenty. While cattle were never held on this scale, most houses had at least a few head. This diminution of livestock is due to several factors. In previous times the animals were allowed to run free and scavenge, but in the past ten years an old law that all livestock must be penned began to be enforced. The recent fencing movement in the village has added to the difficulty of keeping animals. A penned animal must be fed and the pasture must be kept up, all of which costs money which the peasants do not have. Also, the prices of animals have increased at least tenfold in the last thirty years while the peasants' earning power has scarcely kept pace. Animals which used to be kept primarily for their use value - eating - have now become a market commodity. As they were 350 years ago, the small-scale herders have been forced out of beef production.

These problems are small, however, compared to those created by the removal of the forest cover. In the first place, with the increase of population there is, on a per household basis, less forest or *monte* available for use in the slash and burn agricultural cycle. This forest-population density problem has grown more acute now that the entire *campesino* populace can no longer move so freely across the land surface, given the increased amount of fencing. Second, some years ago commercial graziers introduced to the central provinces a new cattle grass known as faragua (*Hyparrhenea rufa*). This drought resistant pasture grass thrives in the savanna. According to Bennett (1968: 62), 'Until faragua became generally available, cattle raising was a poor economic venture in many parts of western Panama where today the industry flourishes.' Faragua appears to dominate other seedings, and where it becomes established the forest cover regenerates only slowly. With the withdrawal of cattle from Los Boquerones faragua has become even more difficult to cope with, and in the dry season it often permits fires to spread out of control, with the result that the resting and growing trees are damaged. Lastly, when the Fábregas left and the concrete highway was completed, thus lowering transportation costs, the peasants took up raising sugar cane as a cash crop. Following seeding, cane takes eighteen months before it is ready to harvest commercially, but then a good field will produce a profitable harvest for four years. Afterwards, in the absence of ploughing, the cane continues to grow, although on a diminished scale, and the forest does not seem to re-establish itself. Thus, due to population expansion, new grasses and the seeding of sugar cane, the amount of forest available for use in the traditional swidden cycle has been decreasing.

The consequences of the changes have been twofold. As a result both of resource depletion and market pull, the people have been turning from production of their own means of subsistence to cash purchases - a reverse form of import substitution. In place of thatch, for example, corrugated roofing now is often purchased; for vines, nails are used; metal eating utensils have replaced those fashioned from wood and gourds. Cash received from the sale of sugar cane has enabled such purchases to be made, but market prices seem to advance faster than earning power.

The changes also have led to a narrowing of the productive base. Cattle, hogs and several other food sources have been nearly eliminated, while the land is yielding smaller harvests. The people say they are working harder now and harvesting less. Given declining productivity the labour embodied in a unit of the traditional crops has increased, but the same cannot be said of harvest volume or market price, for the latter is determined by the large-scale, more highly productive, commercial farmers, whose aim is to lower costs.

Resource depletion, primarily the result of external forces, has had a dramatic impact on the subsistence system. The traditional economic mode of rotating fields and exploiting raw materials was predicated on the existence of sufficient resources. Such an economy is a steady-state system in that it does not use natural wealth at an accelerating rate. So soon, however, as the area became economically accessible, capitalism began to take over some resources and help deplete others. The elements comprising factor push and market pull have begun to undermine the prevailing mode in the interior, and the subsistence system has been pushed to its limits. To be sure, Los Boquerones is still predominantly a subsistence economy, but it is balanced between a final stage of production for use and the early stages of production for exchange.

The anthropological commonplace, that no community can be understood as an entity in itself, is certainly true of Los Boquerones. But it remains insufficient to focus only on the marketplace transactions which the villagers maintain with townsfolk and urbanites of Panama and to isolate the current economy from its history, for the conditions of the present system are written in its history, and this is a history not of the marketplace but of changing productive arrangements. In terms of the metropolitan areas the value of the interior shifted from a source of plunder to a source of raw materials. Then, for a long period the area became marginal and was occupied by people marginal to the predominant mode of production, a people defined as being both within and without this mode. The subsistence economy did not grow up by itself, simply as a reversion to a more 'natural' economy, as the fundamental uniting of the labourer, his labour and his product. The economy of the interior is not only a recluse but a reactionary system in that it was formed and maintained as a reaction to capitalism. Developed in those areas of the country where there was little interest in investment, subsistence - production for use - was an adaptive way of life, a recourse to counter simultaneously potential exploitation by the market and exclusion from it.

From a national standpoint a subsistence system represents a break, non-integration, an island in the economic process. The historical lack of transference of goods or services from countryside to city not only reproduces itself but creates low social and political integration between the two sectors; neither the political nor social order of Panama has ever rested on the support of the peasantry. As the national economy penetrates into the interior, however, new forms of political organization and domination may be expected, a process presaged, perhaps, by the recent evident link between sugar-cane mill ownership and political power.

Thus, the subsistence system is itself a contradiction, a dual economy (Boeke, 1942). Its characteristics are decidedly non-capitalistic, one of its most notable features is an almost complete lack of 'capital' formation. Yet, the system was defined and generated by capitalism. The peasants want very much to partake of consumer goods - 'civilization' - but do not want to sell their subsistence crops for cash. The change, therefore, from producing rice for eating to raising sugar cane for cash is both a radical and a non-radical shift. The peasants desire a better future, a future which would mean escaping from small-scale agriculture. But for them profit is *viva* - a trick of extraction by sharp traders. Farming is not a business, but all desire access to that one resource from which they are excluded, capital funds. The dual economy of Panama, therefore, refers not, as Boeke would have it, to the split between Isthmus and interior, but to the contradiction which is within the countryside itself. Formed in reaction, in opposition, to an external set of conditions, the subsistence economy contains within it those very conditions, its own future and the seeds of its own destruction.

3 Household production: subsistence and surplus

Excluded from the transit area, the farmers of the interior developed a system of non-commodity production. The historical conditions of a small, dispersed rural population using simple productive techniques, having relatively easy access to raw materials, and transmitting few goods to the cities led in the countryside to a system of production for immediate consumption. Huddled in small domestic units, the *campesinos* have through the years produced primarily to meet their own needs and not to augment their wealth. This fundamental orientation - household production for household consumption - is reflected directly in the people's ideology. Production for livelihood, however, does not exclude producing a surplus; the enigma of production for use is not whether a surplus is produced but why the surplus which is produced does not accumulate and transform the system.

Production for use

By labouring, whether in the fields or at home, the individual transforms raw material or an object and creates an item of use or an increment in utility. In Los Boquerones the 'production line' passes from fields to home. Consumed in the home, the object enables the worker to begin the cycle afresh.

Such production for consumption stands opposed to production for exchange in which an object is produced not for its use qualities but for its 'power of purchasing other goods' (Smith, 1910: 25). In itself the exchange item has no utility, no meaning, for the producer; not without reason did some of the older men in the village lament that sugar cane, produced for sale to the mills, 'will be the death of us all'. 'For the self or for others' expresses the reproduction opposition of endogamy versus exogamy as well as the production opposition of use versus exchange.

The qualitative nature of the object dominates in production for use, but quantities also are important, for a good must be produced in sufficient amounts to meet the producer's needs. The *campesinos* are acutely aware of the quantities of product they require. The Los Boquerones system of fabricating useful objects, however, is not predicated upon ever-increasing production. Sufficiency not surfeit is the aim.

In a system based on production for exchange the objects created have an ultimate use; so also in a system founded on use production, exchange may be carried out. The peasants have traditionally not only purchased items on the national market but exchanged products and labour amongst themselves. The exchange sphere of the use value system, however, remains subordinated to the aim of self-sufficiency.

The organization of production

There is in the countryside a conjunction of functions among kinship groups, the household and relations of production. Economic units are dependent upon and reproduced through ties of kinship, while domestic relationships of kinship are production relations. Although agricultural requirements do necessitate some co-operation among households, goods are created primarily within the domestic group. Some households, to be sure, do produce speciality items or provide unique services, but such activities are secondary to the fundamental task of producing one's own sustenance.

Household groups are based upon or modelled after a specified kinship unit, the nuclear family. Not all bonds of kinship, however, are economically important. Family ties inter-link domestic groups within the village and beyond, but such external ties are not fundamentally of a value-producing or consuming nature. They are conceived to be moral bonds which may or may not be deployed for utilitarian purposes. At this level kinship is superstructural to production.

A reciprocal relation links kinship and co-residence. Regardless of their true relations, household members become 'as if' kin, while the relations between truly close kin living in different homes are sometimes denied.[1] Viewed from within, kinship in part arises from and incorporates production relations. It is not accidental that males, the primary producers, have the most freedom to claim or deny ties of parenthood. But even within the household kinship is not completely devoted to or exhaustively defined by the economics of everyday life, for the functions of the family extend beyond production.

Household family bonds are conditioned by the two invariant factors of sex and age. Internally the division of labour separates the members such that males, females and children throughout the village are alike in terms of their work but unlike in that the fruits of their labour become the property of different households. Thus, households resemble one another internally but are economically separate.

The often repeated proverb of the people reflects directly the productive relationships: 'the man is in the fields, the woman is at home.' In a productive sense the domain of domesticity is the female's while the sphere of crop production is the male's. So deeply rooted is this specification that its inverted, untrue form, also becomes a reality: the person who works in the fields - man or woman - is a male. Men, found cooking, and thus reversing the proper order, are said to be under the domination of their spouses. When asked why he carried his instrument of production - the machete - even when not intending to use it, one man replied: 'I am not a woman.'

Households cannot and do not exist if the fundamental tasks are not carried out in co-operation. There is a veritable flow of goods from fields to house, from man to woman to consumption. Rice is seeded, cared for, harvested and brought to the home by the man: the woman dries, shells, cleans and cooks it. Withdrawal of the functions destroys the unit.

Rooted in the productive system, the division of the sexes is marked in all spheres of the social life. Women are linked to the home and family, while men are said to be 'of the street'. After work men often visit others, wander about, or sit at a store conversing; the man who always stays at home is *pendejo* - under the domination of his spouse. Women, on the other hand, also visit one another, often taking their children with them, but the female who persistently wanders about alone is known as a 'woman of the street', a prostitute. On ritual days men are prohibited from working in the fields and are brought into the sanctity of the home. But as if in reaction to this equation of themselves with domesticity, femininity and sacredness, they spend their time in the *fiesta* of the street. The solitary agriculturalist of the ordinary workday expresses his communion with others on strictly social and religious occasions, activities which do not undermine the postulate of individual production (Gudeman, 1976b). By contrast to this 'horizontal integration' of the male which is linked to his productive role, the woman is the focal point of 'vertical' integration or reproduction through the central, unbreakable tie she forms with her children.

The essential economic activities of the household, thus, are founded on a sexual division which itself is internal to kinship units. Kinship, residence and production at the household level are indissolubly linked.

The second elemental division in the household is founded on age. Differentiated from animals but starting from an almost nonhuman status, children are 'raised' (*criar*), or integrated into the household, and trained to undertake sex-linked activities. But as soon as the youngsters become capable of fulfilling their sex-linked roles - when the adults have been reproduced - they leave the unit to form their own. In fact, in the period immediately preceding fission, a home exhibits tension, for the internal relationships are not founded on an equivalent exchange. Senior members attempt to keep junior members as labourers in and for the household, since the value the latter produce exceeds their maintenance cost; but junior members want to gain total control over their labour and its products. This conflict, generated by household productive relations, leads not only to their dissolution but also to their replication over time.

The people's expression 'to get an interest' captures the essence of the relationship between the generations and is an important formula in its own right. Getting an interest is a mechanism for dividing goods and rights to labour. Since it is males who work in the sphere of agricultural production, the rule pertains mostly to them. Before a child becomes a productive member of the household, small quantities of goods, of utilities, are placed in his name - chickens, hogs, even cattle. The child should assume some responsibility for maintenance of the valuable. Although ultimately used for household consumption, the good is recognized as pertaining to the youngster. Later, when a male child begins to work productively, his contribution is recognized in a more direct fashion. When the child works for his father, either he receives some of the cash earned by the sale of an agricultural product or a portion of a field is placed in his name. Although in the latter case the goods of the field flow to the household, rights to the instrument of production, the field, become the child's when he leaves the home. Thus, a youngster gains an interest in working as he earns an interest in the reward. If a young male works outside the household for cash, he divides his earnings into two portions. One, given over to his father, is used for the household and indirectly his own maintenance. The other he keeps for personal use. The size of the cash shares is negotiable; indeed, it is the subject of discussion and often a cause of friction between father and son.

A father tries to keep his son at home by assuring him of basic subsistence and providing him with ever greater shares of the household product. His very actions, however, generate the desire to leave, and the more of the product he returns to his son the less valuable is the latter's labour to the father. The son, for his part, has in his father's home the constant assurance of maintenance, but, the older he grows and the more he works, the more productive he becomes and the more onerous the fetter of extraction. Arguments flare up within the household until the son leaves, builds his own home and subsists on his own, taking with him, it may be added, some portion of accumulated value from his natal household.

The position of the growing female is rather different, for the young girl works in the home with her mother. She, too, leaves her natal family upon maturity but is not so directly linked to property, in part because her labour - however useful in the home - is a service and cannot within the subsistence system be transformed into production goods. Unlike a male, when a female leaves she moves directly from her parents' house to that of her spouse. By this sudden transition, however, she becomes mistress over her own household as opposed to being under the domestic domination of her mother.

A family does not provide an estate, an inheritance, or income-yielding property when a child leaves. What it furnishes, rather, is the stock of goods - land, seed, and some consumption items - so that a son can sustain himself and build up his own production process. The family's own stock is reduced by this amount, but it also has one less worker-consumer to support. The household family provides its departing members with 'working capital', not a fixed investment.

By moving into a separate home a new conjugal pair reproduce the conditions of production, domesticity, and the family. The developmental cycle of kinship units presents the conditions for the reproduction of the economy, yet it is the economic implications of the kinship bonds - the complementary nature of the conjugal tie and the growing extraction of labour by parents from their children - which give impetus to the breakup and reproduction of nuclear families.

Little real property is accumulated and passed between the generations upon death. The household depends upon live labour, and it is the conditions of the live labourer which it continually reproduces. Although households as economic units are not corporate and continuous, their reproduction is assured by their kinship basis.

A subsistence ideology

Items of the national market have, for the peasants, a 'value' - 'civilization' - over and above their utility, but products of use emanating from the countryside also have a symbolic import. Rice not only provides sustenance, it represents the investment of past labour and the efficacious functioning of a household. The subsistence system is not bereft of an ideology.

The word subsistence has been variously employed in the literature, but at worst my usage has been Janus-faced. On the one hand, and from an external standpoint, subsistence refers to self-sufficiency, the production of objects for use. Since the subsistence factor underlies every economy, its sharpest meaning emerges when contrasted to production for exchange. The person who produces for self-consumption is a subsistence producer. It is primarily a question of dominance, of the gearing of the system toward one or the other alternative.

On the other hand, and from an internal standpoint, subsistence refers to the standard of living, to expectations of the folk about their own maintenance or consumption. In Los Boquerones this facet is expressed ideologically as 'necessity', what the person needs to live. But like the Italian deity the two faces of subsistence are related. In thought, one produces because one has needs, although in fact one may 'need' what one produces. Production for the household creates the consumption requirements of the household.

For the peasants necessity is a cultural idea about the natural needs of humans. A man, when asked to explain a past behaviour, often raises his hand and motions toward his mouth, adding 'The life of one is this.' The foods rice, maize or bread, cooking oil, beans and some meat are necessities. Clothing and housing are necessities as is the faculty to obtain aid to cure an illness. Money in itself is not said to be a necessity; rather, it is required only in so far as it can procure items of use. As many pointed out, they work not to earn money but to have food - 'to live, nothing more'.

> We think of working in order to eat. I want to have food to eat. We don't make sums of what we gain or lose - no one thinks of this. We think about getting food to eat, of having enough for next year. I work hard now to have enough food for next year.

The consumption requirements of the household vary by family size, by the number of consumers. In monetary terms most families in Los Boquerones could subsist on $4.00 to $8.00 a week, depending on household size. Given me by a selection of persons, these calculations exclude all but a minimum of 'luxuries'; they are a true subsistence level. But subsistence - necessity - is a cultural, not a natural, norm, and it should not pass unremarked that there is fairly wide agreement about this standard. The people share a set of consumption desires and expectations, stated either in monetary terms or in quantities of objects. For a certain range of goods, demand is inelastic, to use a different but not inappropriate terminology. But this points to a second important aspect of subsistence as necessity. Through time, demand is relatively stable, compared with capitalistic countries. Of course, the standards of necessity probably do change and probably have changed over the years; certainly, they vary spatially among the rural folk in the interior. But the production of objects for subsistence is antipathetic to rapidly enlarging standards of consumption, to the expansion of production possibilities and to outward moving demand curves. The interior itself is not an area of rising consumption expectations, and of course this empirical characteristic of the present day is a result of the historical conditions in which the system was created.

Necessities stand opposed to both vices (*vicios*) and luxuries (*lujos*). Although the concept of vice is an integral part of Roman Catholic theology - dealt with extensively, for example, by Saint Thomas Aquinas - in the countryside it has a fundamental economic resonance. Vices are both categories of goods and necessities carried to extremes. Tobacco, alcohol, and the lottery are vices; coffee, enjoyed by all, if neither vice nor necessity is at least a 'taste', a pleasure, a slight indulgence. On the other hand, meat - given that beans are available - is a necessity on occasion but a vice when often demanded. Vice, then, is not a religious concept conveying eternal truths about the universe; vices are not in themselves bad for the soul. Rather, vices, which can ensnare people and lead to compulsions, may undermine the fundamental task of gaining subsistence.

Vice and luxury are opposed to necessity as production for exchange is to production for use. Goods defined as necessaries generally include only items produced in the countryside, although not all locally produced goods are necessaries. Items classed as vices usually must be purchased, while luxury goods - shoes, household furniture, better clothes - are all obtained with money, usually gained by the sale of sugar cane. But this division of items is not the expression of a categorical separation of goods into 'spheres of exchange'. The categories are not conceived in a vacuum and then translated to economic action. Market goods are not morally bad: rather, the juxtaposition of the two economies produces both the division and the contents of the categories.[2] Maintenance precedes indulgence.

Throughout the central area of Panama, rice is the staple food. This production fact has been transformed by the people into a consumption demand, a demand which from the standpoint of the market appears to be 'irrational'. The people say that they cannot live without rice, that a meal is not food without rice, and that the first food they buy when needed is rice. As the people express it, consumption, the demand schedule, determines production. When the peasants are out of food and go to the marketplace, no other product can substitute for rice, regardless of relative prices.

Given the growing of rice as the essential condition and sign of the self-reproducing system, its production has become a source of pride. The work is hard, but men take pleasure in a rice field they have carefully weeded; they will visit and revisit their fields throughout the growing season, ostensibly for important reasons but also for pleasure. The female tasks of shelling, cleaning and cooking rice are arduous and time-consuming but also satisfying.

Beyond this pride of workmanship, a loft full of rice, drying above the kitchen fire, provides a sense of security, and men will admit they are embarrassed if they have no rice stored in the home. When reciting the changes of the past thirty years, the peasants often say they have less food now. Their actual food intake may have declined but the statement also reverberates with the idea that with more cash and more market purchases, homes are less and less the physical repositories of the food supply. Given the traditional satisfactions, the change is viewed as a loss.

Rice is sold only to meet a necessity, such as buying other food or medicine, or paying labourers to raise more food in one's fields. The cash received for selling sugar cane, however, is an 'abundance' and expended on a greater range of goods. Similarly, when in need of rice a man will seldom ask another to sell him some from the family storehouse, for he knows that everyone else has also raised for domestic use. But he would also be embarrassed to ask, for his own fundamental sense of respect (*respeto*) is founded on being independent of others.

In Los Boquerones, as in many Hispanic cultures, having and maintaining the respect of others is a central goal; however, this ideological configuration has, in the countryside, an economic foundation. A man's first responsibility, his fundamental obligation (*compromiso*), is to maintain his family and finish the work he has commenced. From this flows all the detailed tasks he owes to the household. To be able to defend the self (*defenderse*), to be independent and self-sufficient, is the requisite for gaining the respect of others. Outside the household, one avoids *compromisos*, being dependent on others, being in debt; and a person bears little responsibility for others in the community. The people will discuss in great detail their sicknesses, for which they are not thought to be responsible, but never their hunger for which they are. When returning from the fields a man will exclaim that he is *estropeado*, worn down, that he is tired after accomplishing the work of today, but not *cansado*, too tired or fundamentally weak to work tomorrow. To have no rice, to be a salaried worker in agriculture brings a loss of respect. For men a source of pride is being able to order oneself in the fields and not having to work for cash. When one's own work is finished, then earning money or raising sugar cane takes on a different complexion. It becomes something extra, an abundance. Market goods are attractive but not bound up with the fundamental definition of the individual within the community.

The social relations of production, the necessities imposed by living in a household, lead rather directly to the economic goals of the individual. The ideology extols neither the generosity and power of a sultan nor the economic progress of an Horatio Alger. The man of respect, the person who can defend himself, maintains a household and leads a passable existence. His pride is in his repetitive work rather than accumulation in production.

If 'accumulate, accumulate' (Marx, 1967a: 595) is the battle cry of a growing capitalism, then 'economize, economize' is the rally of the countryside. 'Economy' does not for the peasants mean institutions, exchanges, a sector of society. Economy means economizing, cutting costs, and that in a double sense: avoid money costs in agriculture for subsistence and minimize the costs of the household. Economizing costs does not create profits, but it is the converse of being consumed by vices, and with enough economy, savings for the future will be created.

The countryside, in the people's view, does not generate capital, meaning funds for investment. Lacking capital the peasants are unable to obtain outside capital, for government agencies and commercial banks require that a debtor have a guarantor or collateral to stand behind an agricultural loan. On the other hand, the people also do not want the obligation, the risk and the danger of a long-term debt to an outsider. Production for use is both a protection against and a reaction to exclusion from market capital.

Yet, the interior is sited on the borders of capitalism, and the peasants are caught within the contradiction of being constrained to live within the subsistence system while being aware of other, more powerful possibilities. Even more, they are defined as the residual members of this larger system - the ones 'lacking civilization', a term meaning backwardness in a spatial, historical and economic sense. Thus, their aspirations are in fact dual, almost contradictory. Beyond securing subsistence, everyone would like to be rich, but there is no notion of transforming the countryside, of raising agriculture to a business. It is in the niches of society that one has hopes for one's self or one's children: working as rural school teachers, mechanics, dressmakers, truck drivers or as aids to local mayors. To be salaried in the broader economy, in contrast to agriculture, is a desire. Even these grander aspirations, however, have a certain duality. On the one hand, the peasants see themselves as joining the labour force, becoming salaried workers. On the other, to own a store in the countryside or a town is also an aim. Both goals are related to their notion of profit.

'Profit is a mystery, we don't understand it' were the words of a perceptive informant. The peasants may be mystified by the origin of profit, but they do have an ideology concerning its locale. Profit - *ganancia* - is a thing of the marketplace and exchange. Not an accounting concept about production, based on the difference between a stream of revenues and costs, profit is made by the *negociantes*, the merchants and dealers. It is perceived that subsistence goods which are sold for a low price must be purchased for a high one. When raised in the home, chickens are not a source of profit, but if they can be bought for a low price and sold for a higher one, they are profitable. Selling only when in need of immediate cash and not encountering a free market, the peasants find themselves at the mercy of the purchaser. Their real life experience itself leads to a trading conception of profits and the notion that merchants are robbers. Profit is *viva*, a trick of the sharp individual. But this understanding is consistent with the internal system, for the goods sold to local traders - rice, maize, yuca, animals - are not sources of profit in the countryside, since they are raised for household use and in some instances are, in a book-keeping sense, losing ventures.

This same view is carried over to the cash crop. The raw cane does yield, it is thought, a profit, but the only difference between the sugar-cane mills and the local traders is that the former 'rob scientifically', getting rich on the sweat of the peasants. In fact, so large are the mills, it is said that they do not even need to use the money of others to finance their business, being almost a bank in themselves. Money is the fount of profits, and goods are valuable (*valen*) only when in the hands of the rich. The peasants, excluded in their view from the true source of profits, are quite logical in their aspirations; they can only hope to participate in the larger economy as salaried workers or as petty storekeepers.[3]

If the subsistence system is not in want of a secular ideology, it also does not lack a theology. But this is not the land of Calvin. Resignation and restoration are the keynotes. Abstinence in order to accumulate is neither a virtue nor a sign of the elect.

A good head and hard work, even a 'good heart', may not themselves lead to an easier life. Economic fortune is ultimately a matter of *suerte* - one's lot or fate. To win at the lottery, to have cattle which regularly produce calves, to have a field yield well, to be served a large portion of meat, to be bitten by a snake, or to obtain a large loan from the sugar mill, all are matters of good and ill fortune. Ultimately, they are functions of one's destiny which God has laid forth.

It is too facile to say, however, that a religion of destiny promotes only *post hoc* explanations, for this same theology requires that the individual 'struggle ahead', to meet fortune as it comes. The religious beliefs may not exalt the entrepreneur, the budding capitalist, or even special personal qualities, but they also do not allow the person to be a laggard, for he expresses his faith through acts, through labour. The man who stays at home, who lies in his hammock, will never be helped by God, and may even be condemned.

Indeed, earth is the battleground between God and the devil, good and evil. The devil constantly presents temptations to lead one from the path of good. Some, it is said, have even made pacts with him, exchanging their souls for worldly riches. And the wealthy, if not now aligned with the devil, certainly walk in his path. God only helps one to persist and endure (*pasar*). This ethic of poverty does not lead the people to abjure their earthly gains; rather, it is a consolation, perhaps a justification, for their lives.

The Christ figure is a symbol of death and defeat more than resurrection and rebirth. And the saints have assumed roles of unusual prominence, for it is to the saints, and only to the saints, that one can make vows asking for help on earth. Saints are appealed to in times of distress, primarily to restore the prior order or to ameliorate a current problem. They are not asked to help in those contexts where man himself is thought to be responsible. Saints do not rid one of hunger pangs, but they can ask God to send away an illness or crop plague. Saints are not appealed to in order to have a better or different future, for that is part of one's destiny.

To look to the religion as the cause of economic 'stagnation' in the interior, as the rationale for the subsistence system and its complex relation to capitalism, would be a distortion and simplification not warranted by the historical facts. Yet, it is true that this ethic of rising each day and facing one's fortune by working the earth, this theology linking the devil and wealth, is a non-revolutionary, non-reforming set of beliefs well suited to a perduring subsistence economy located on the margins of capitalism.

Subsistence and surplus: theory

If, for the people, necessity is a certain level of living, what determines this standard? What factors set the return to the worker? Posed thus, the problem explodes into a galaxy of further questions. If there is subsistence, is there also surplus, and what is the form of the surplus? Further, what determines the division - the distribution - between subsistence and surplus? Then, granted the existence of both subsistence and surplus, how can they be measured? How may a 'heterogeneous' collection of use objects be measured in a non-monetary economy?

Cross-cutting these apparently empirical questions are the approaches of the different traditions in economics. If the semantics of the words 'subsistence' and 'surplus' have been the focus of contention among anthropologists, much the same is true of economists (Dalton, 1960; Harris, 1959; Keyder, 1975; Mandel, 1968; Pearson, 1957). Therefore, a slight diversion from the 'facts' of peasant life seems appropriate here.

In neoclassical, marginal models the term 'surplus' has no single meaning nor precise analytical status. Surplus might refer to the retained earnings of a firm (assets in excess of debt obligations), or to profit, or to the excess over expected profits, as in windfall gains secured through international currency transactions. In the context of the marketplace, however, the word is devoid of content, for so long as the free market obtains, as long as the 'invisible hand' is operating, then prices will always be the result of the adjustment between supply and demand, and there can be no surplus of goods, no excess of supply (leaving to one side the Keynesian problem of 'effective demand').

The concept of 'subsistence' occupies an equally ambiguous position; it, too, has no theoretical status. Sometimes the notion of the subsistence producer is reduced to an *ad hoc* list of empirical characteristics (Wharton, 1969, 1971: 153). A subsistence level then becomes that vague entity defined as a standard of living above the bare minimum but 'controlled by economic-socio-cultural factors' (Wharton, 1971: 161).

On occasion, however, the marginal theory of the firm is applied to subsistence farmers in order to bring them within the orbit of neoclassical economics. Many of these models are built upon the original work of Chayanov (1966). Nakajima, for example, provides a sophisticated series of models which might well have been elaborated by a latter-day Chayanov: 'For the family farm in equilibrium the "marginal productivity of labor" equals the "marginal valuation of family labor". . . . The amount of output . . . is determined by the production function' (Nakajima, 1969: 169).[4] Many of the same criticisms, however, that were levelled fifty years ago against Chayanov could aptly be applied against these models: the problem of the relation between model and facts, the assumption of equilibrium or unchanging output through time, the isolating of the farmer from the broader economic system of which he is a part, and the need to postulate subjective utility.

Opposed to the neoclassical approach is the specifically Marxian version of subsistence and surplus. For Marx surplus arises from the difference, which the capitalist is able to exploit, between the individual's labour and his 'labour power'. Labour designates that which the worker 'incorporates' in an object or expends in production; labour power refers to the capacity of the human for labouring, to the individual poised or ready to act - the 'aggregate of those mental and physical capabilities existing in a human being which he exercises whenever he produces a use-value' (Marx, 1967a: 167). The capitalist purchases the worker's labour power but uses his labour. Herein, according to Marx, lies the origin of profit.

How may this surplus, this difference, be measured? On the one hand, human labour, labour expended, may be evaluated simply in time units; the labour value of a product, thus, consists of the total labour time required to produce it. But evaluating labour power, or the commodity the worker sells to his employer, poses special difficulties, for how can the capacity of the human be measured in terms commensurate with those used to calculate the value of his product? Marx's well-known solution to this puzzle was to suggest that the value of labour power corresponds to the labour time required for *its* production, the labour time it takes to maintain or reproduce the labourer and his family. The worth of labour power is the labour time necessary to produce the means of subsistence for the labourer. Therefore, surplus designates the difference between the labour 'embodied' in a product and the labour 'embodied' in the subsistence goods which the labourer consumes (Marx, 1967b: 785–6):

> If man were not capable of producing in one working-day more means of subsistence, which signifies in the strictest sense more agricultural products than every labourer needs for his own reproduction . . . then one could not speak at all either of surplus-product or surplus-value. An agricultural labour productivity exceeding the individual requirements of the labourer is the basis of all societies.

Marx's conception, however, raises two immediate problems. First, it is not totally clear to what degree the concepts which he worked out for the analysis of capitalism apply directly to non-capitalist societies. In the Marxian framework the concepts of subsistence and surplus are closely linked to those of exploitation and the relations of production. Exploitation is the extraction of surplus, and Marx accords a decisive role within a mode of production to unpaid surplus labour and the form of exploitation (1967b: 791).

But we may also ask whether a surplus can be produced 'outside' a relation of exploitation.[5] If, for example, it is held that surplus arises only within the context of extraction, then the concept when applied to the historical conditions of Los Boquerones and the Panamanian countryside encounters immediate difficulties. For several hundred years the peasants paid either no or a very small labour rent. In Los Boquerones the rent was increased during this century and was paid in cash or labour up to the early 1950s. Since then no rent has been paid, but the impending purchase of the land will amount to a capitalized rent, at a high rate of extraction. Thus, in Marxian terms the people have moved from producing no surplus to a small one, to a larger one, to no surplus, to the possibility of having to produce a very large one. Yet these shifts in surplus extraction have not been accompanied by any change in the essential relations of production, nor have the people's level of living and total output changed much in consequence of the different degrees of exploitation. Are we forced into the position of saying that surplus sometimes exists and sometimes does not in Los Boquerones, that surplus is 'created' by extraction and not by the production process itself? How can we account for the peasants' productive system in terms which are theoretically consistent for this time period? Without question, the overall nature of the relation between the peasants and urbanites has had a decisive effect on the pattern of the rural economy, but the relation itself does not explain much about the actual level of product output and labour input.

A slightly different but co-ordinate view of surplus has been provided by Piero Sraffa, and by following dimensions of his approach we may be enabled to bring the analysis of a system of production for use back into an encompassing economic theory. Sraffa's work represents, in certain ways, a reinterpretation of and return to Ricardo's first or corn theory of profit. In his Introduction to Ricardo's works, Sraffa (1951: xxxi) points out what was 'never explicitly stated by Ricardo' that

> in agriculture the same commodity, namely corn, forms both the capital (conceived as composed of the subsistence necessary for workers) and the product, so that the determination of profit by the difference between total product and capital advanced, and also the determination of the ratio of this profit to the capital, is done directly between quantities of corn without any question of valuation.

In other words, and more broadly, the calculation of profit - or surplus - and of the rate of profit - or rate of surplus - may be made for commodities which 'produce' themselves without having to make recourse to pricing mechanisms, and without having to drag in the notion of exploitation.

In his own work, published nearly ten years later, Sraffa essentially considers surplus to be the excess of production over material input. The return to capitalists and to workers is a division of this surplus. In the Sraffa system wages are not 'advanced' to workers as subsistence before production, as payment for their labour power, or as an 'input cost'; rather, salaries are 'paid *post factum* as a share of the annual product', although Sraffa also suggests that wages could themselves be divided into necessaries and a share of the surplus (1960: 10). Thus, where for Marx the return to the worker is subsistence and to the capitalist is surplus, by the Sraffa method surplus may be considered to be the excess of output over input for a commodity which is used to produce itself, such as corn. This surplus is 'then' divided into wages and profit portions.

In fact, the Sraffa system is not at odds with the Marxian version of surplus, and this brings us to the second problem facing Marx's conception of surplus - a problem which has its precursor in Ricardo and is left undetermined by Sraffa - namely, the problem of distribution. What determines in capitalism the division of output between worker and capitalist? Specifically, what determines the level of the means of subsistence of the labourer? In the context of a subsistence system we may frame this as what determines the division of the output between, on the one hand, the necessaries which the labourer consumes in order to produce and, on the other, the goods which he devotes to other purposes? The neoclassicist, of course, would state that such questions of distribution are answered in the marketplace; but even leaving to one side the theoretical attacks launched against this theory by Marxists and neo-Ricardians, there remains the simple fact that a pure subsistence system has no marketplace where the relative returns to the different 'factors of production' could be set.

With respect to a 'subsistence level' Ricardo, Marx and Sraffa all allude to cultural elements entering into its determination. Thus, Ricardo (1951: 96-7), taking note of Torrens's observation that India and Russia had 'different habits of living', stated:

> It is not to be understood that the natural price of labour, estimated even in food and necessaries, is absolutely fixed and constant. It varies at different times in different countries. It essentially depends on the habits and customs of the people.

Later, Marx was to speak of the 'so-called necessary wants' which are the 'product of historical development', and he went on to argue that 'there enters into the determination of the value of labour-power a historical and moral element' (Marx, 1967a: 171). Shades of the same argument, though with a different twist, also can be found in Sraffa (1960: 33).

Taken at face value, there appears to be a certain amount of question-begging here, and with a subsistence system - a classless society which grew up on its own - the infinite regression, the explanation by way of 'history', reaches its terminus. On the positive side, however, this way of approaching distribution brings in the socio-cultural context at the outset. Distribution does not appear 'later' as an outcome of market pricing of produced objects; it is inserted into the production process itself.

Perhaps the problem of seeking the determinants of a subsistence level can be understood better by contrasting, in a broad fashion, the two approaches in economics used to explain value. For the neoclassicist value is not an absolute or intrinsic quality of an object; rather, it denotes a relation between objects. Value is the rate, the price, at which two things are exchanged; and the value of any object clearly changes with any changes in the values of other commodities. In the neoclassical tradition value is a relational concept.

By contrast, the Ricardian tradition, by basing value upon labour, appears to begin with an absolute definition of value. For Marx the value of a good or person is the labour time socially necessary for its or his production, a value which varies only with changes in productivity. The value of the labourer is the labour time needed to produce his subsistence.

But it could be argued that within this second tradition there is also a relational notion of value, though of a different order from the neoclassical idea. The crucial element may not be the actual subsistence bundle of goods which goes to the labourer, and thus determines his value, so much as the size of his share of the output compared to that which accrues to the capitalist. The entity to focus on is not the subsistence level in itself but the subsistence portion in relation to the surplus, using the latter term in either the Marxian or Sraffa sense. As Joan Robinson (1965b: 33-4) has cogently pointed out, the 'rate of exploitation' (the ratio of net profits to wages) varies by different cultures, even among capitalistic nations participating in the same world market. The essential problem, thus, is to grasp the determining factors of distribution without reducing these to purely 'economic forces' (marginal products) or to some presupposed standard of living of the labourers.

Subsistence and surplus: production facts

How do these concepts apply to the real world of subsistence producers? To begin, in what respects is a surplus produced in the Panamanian countryside?

Surplus, as I shall use the term, refers to the difference between the object produced and the costs which are required to produce it (seed, equipment, and necessary maintenance of the worker, *while*

engaged in the production process). The disparity between this definition of surplus and Sraffa's is relatively minor. Where I treat the labourer's maintenance for time of production as a set input cost, for Sraffa the whole of the wage is variable and is part of the surplus.

In the villager's economy rice and maize are used to produce rice and maize. The commodities feed into themselves as both seed and worker maintenance. But the crops not only 'reproduce' themselves, they yield a surplus. To take a simplified example, but one not far removed from reality, in 90 days a man can produce enough agricultural crops for himself and his family to last 365 days. If he only reproduced himself and his family, then with 90 days of labour he would produce only 90 days of food, or he would need to work 365 days to produce 365 days of food. But the rural Panamanians, like most humans, can create in a day more than they consume in a day; or, phrased differently, they can consume in a day less than they create in a day. This difference between output and input represents the surplus created in the production process.

We can measure this surplus in various ways, *but without making reference to pricing mechanisms.* If, for simplicity and only for the moment, we omit the minor costs of seed and equipment, then in labour time the surplus is equivalent to the labour days required to create the product less the labour days needed to create the consumption goods of the worker and his household which they consume during the production period. In abbreviated form:

$$\text{Surplus in labour days} = \begin{array}{l}\text{Labour value of product—}\\ \text{Labour value of subsistence}\\ \text{needed during production}\end{array}$$

Using the above example, 365 days of food are created with 90 days of labour; therefore, 1 day of food is produced by 90/365 labour days, and 90 days of food are yielded by 90 × 90/365 labour days. Thus,

$$\text{Surplus} = 90 - 90 \times 90/365 = 67.81 \text{ labour days.}$$

Similarly, the surplus may be calculated in food days:

$$\text{Surplus in food days} = \begin{array}{l}\text{Food days product will support —}\\ \text{Food days necessary for production}\end{array}$$

$$\text{Surplus} = 365 - 90 = 275 \text{ food days}$$

The surplus also may be measured in food units, a method I shall later use.

Given this definition of surplus, we may designate the rate of surplus formation as the ratio of the output less the input to the input, or - what is the same thing - as the ratio of the surplus to worker's consumption plus seed and equipment costs.

$$\text{Rate of surplus} = \frac{\text{Surplus}}{\text{Necessary maintenance} + \text{Seed} + \text{Equipment}}$$

In the example I have used, omitting seed and equipment:

$$\text{Rate of surplus} = \frac{67.81 \text{ labour days}}{22.19 \text{ labour days}} = \frac{275 \text{ food days}}{90 \text{ food days}} = 3.06$$

But let me insert a speculation here. In capitalism, the rate of profit refers to the ratio between net profit and investment or capital costs. Suppose that a factory, firm or whatever used no equipment lasting longer than a year, then costs would consist of wages plus raw materials, and profits would equal total output less costs. The rate of profit would then be the ratio of profit to wages plus materials. Curiously, these restrictive conditions almost hold for the subsistence economy. To be sure, the peasants do use some steel-tipped tools which last longer than a year, and they formerly paid a small land rent, but if we disregard these minimal costs, it appears that what I have termed the rate of surplus in a subsistence economy corresponds roughly to the rate of profit in a capitalistic economy.

But this brings us directly back to the problems posed in the prior section. What determines the subsistence level, the division of the output between subsistence and surplus, and the rate of surplus?

To begin we can set some practical and technical limits on the range of values which the rate of surplus can assume. These values depend (in the mathematical sense) on the level of household consumption. On the one hand, given the technical factors of production, decreasing the volume of household consumption during the production period leads to a rise in surplus. If the total product yielded by a set amount of labour remains constant, then lowering the amount of goods consumed during production decreases necessary maintenance and increases the surplus. But in practice there is a lower limit to the subsistence level, aside from bio-physical requirements. Because there is no need to spread the product beyond one agricultural year or 365 days, the lower limit of the subsistence level is determined by dividing the total product (again omitting seed costs) by 365 days, to find a minimum product per day allocation, and then multiplying by the number of production days:

$$\text{Lower limit of subsistence level} = \text{Production days} \times \frac{\text{Total product}}{\text{365 Days}}$$

Increasing the surplus by constricting wants means that the individual is making himself support his own unproductive time.

Conversely, if the consumption level is increased during the production period, the surplus drops. But there is also an upper limit to the subsistence level, for if the total product is consumed in a smaller number of days than it takes to create it, the system will not 'reproduce' itself; outside inputs must be secured to sustain it. Thus:

$$\text{Upper limit of subsistence level} = \text{Production days} \times \frac{\text{Total product}}{\text{Production days}}$$

By increasing the value of labour power, that is the amount of food which the worker and his family consume, the surplus drops. When the surplus dissolves into necessary maintenance, the system will reproduce itself, but to the extent that the number of production days is less than 365, other forms of sustenance will be required in the year.

These two technical formulae, setting the limits of the subsistence level, may be substituted into the formula for the rate of surplus, to define the range of values which it may take. When the subsistence level is at a minimum, the surplus is maximized and so also is the rate of surplus. The reverse relation holds for the maximum subsistence level. If again we omit equipment and seed costs, then for a given set of production conditions, the following statement holds:

$$\frac{\text{Surplus}}{\text{Production days} \times \dfrac{\text{Total product}}{\text{Production days}}} \leqslant \text{Rate of surplus} \leqslant \frac{\text{Surplus}}{\text{Production days} \times \dfrac{\text{Total product}}{\text{365 days}}}$$

This reduces to:

$$0 \leqslant \text{Rate of surplus} \leqslant \frac{365}{\text{Production days}} - 1$$

When the entire product is used evenly over 365 days, as it often is, then the rate of surplus - remembering that necessary maintenance is defined as consumption only during the production period - is determined by the number of production days, or the labour input. Of course, this is a common-sense observation in that equipment costs have been excluded, but the calculation also provides a cross-cultural or comparative measure of the functioning of the economy, a measure which is independent of prices.

No system works exactly as specified, and we may consider what happens under two conditions of variation. A change in productivity will, of course, lead to a change in the rate of surplus. A drop in land fertility, for example, will mean that the same labour input will yield a smaller quantity of product, or a larger input will be required to produce the same output as previously. Under these conditions the rate of surplus will fall.[6] This is exactly what has happened in Los Boquerones during the last fifteen years, and this way of looking at the real-life change lends credence to the people's lament: 'we work harder now and produce less'.

If, to the contrary, productivity remains constant, then an increase of labour input will increase total output at a constant rate. One assumes here no product decline at the 'extensive margin' of land, a reasonable assumption given the traditional conditions of the subsistence system. In this situation, if the product is still spread evenly over 365 days, then the overall and per day volume of necessary maintenance will rise but the absolute surplus will peak at 182–3 days and the rate of surplus will drop.

If, with constant productivity the rise in labour input is not followed by a daily increase in necessary maintenance, if with an input rise, the return to the labourer is kept at the same per day basis as previously, the result will be a greater output, a larger overall necessary maintenance, and a larger surplus but at the same rate of surplus. The rate of surplus would be the following where x represents the added production days:[7]

$$\frac{365 + \dfrac{\text{Total production}}{\text{Production days}}\, x}{\text{Production days} + x} - 1$$

To use my previous example of 90 days labour and 365 days of food, the formula would be:

$$\frac{365 + 4.06x}{90 + x} - 1$$

If 10 days' labour are added, making 100 days of labour input, the rate of surplus remains 3.06. The rate of surplus stays the same, but total surplus, labour input and overall subsistence increase.

These final variations bring to the fore some crucial features of a subsistence economy. Given no marginal decline in output, why do not the people increase their labour input, their total output, and hence their daily consumption or their surplus? We are brought directly back to the distributional problem. Specifying the limiting relation between the subsistence level and the rate of surplus is not to argue for a causative link. In particular, we must ask, is the subsistence level and labour input set and the surplus the residual, as implied by Ricardo and Marx? In the context of capitalism this would be roughly equivalent to stating that the wage level is first determined and profits are the 'leavings of wages'. Conversely, is the surplus predetermined and the subsistence level residual? Again, in a capitalistic situation this would be equivalent to positing that the 'normal' rate of profit determines the volume of profit, and wages are the 'leavings of profit'. Or, third, is the crucial determinant the ratio between the two? Is the subsistence level, calculated as daily consumption times the number of production days, determined not in isolation but only in relation to the surplus? Again, in our comparative frame this would be the same as arguing that the crucial determinant is the 'rate of exploitation' (profit divided by wages).

(The reader may here note that my rate of surplus - surplus/input - is, in the simplified version of the preceding pages, a rate of exploitation - surplus/necessary maintenance. This follows precisely because I have eliminated equipment and seed costs from the denominator, but these restrictive assumptions will be partially lifted in the following chapter.)

Surplus, savings and the relations between production units

To grasp better the determinants of distribution we must, I suggest, follow the third line of argument and view the subsistence share in relation to the surplus portion. What happens to the surplus in Los Boquerones? How does it 'manifest' itself? And, given the existence of surplus why is it not augmented and utilized to make the system grow? More broadly, what is the 'extra-economic' or institutional framework which determines the specifically economic facts?

In capitalism the surplus is held by the shareholders of a corporation, as dividends and retained earnings, that is newly invested capital. A single concept, profit, unites the two portions. In Los Boquerones, by contrast, the surplus is divided up and used for different purposes. Many calls are placed upon it.

Surplus 'appears' initially as product (see Figure 1). One portion of this surplus product is the surfeit when production exceeds yearly consumption. This surfeit is converted to savings or is used to make market purchases.

In addition to their ideas about necessity, the people have the concept and practice of saving (*ahorros*). When yearly production exceeds yearly consumption plus purchases, then a savings, a surfeit of actual objects is created. To the contrary, if yearly consumption exceeds yearly production, a household incurs a deficit or dis-savings. Savings, like subsistence, is a concept linked to time, but unlike necessities which may be calculated on a daily basis, savings are only augmented or diminished when the agricultural cycle has been completed.

FIGURE 1 *Subsistence, surplus, consumption and surfeit*

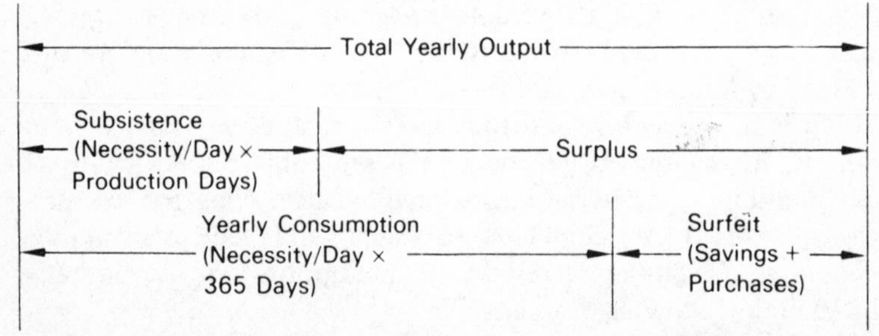

Since savings are that fraction of production above consumption, they are in part a function of self-abstinence. But unlike the capitalist who, by 'abstaining' from consumption raises his capital funds (dividends versus retained earnings), the peasant farmer by abstaining from consumption creates only true savings. For the people, deferred consumption is not investment, it remains a reservoir of value. A surfeit of rice which is not consumed is only converted - to chickens, hogs, cattle or sometimes a house. The peasants are very explicit, for example, that chickens kept about the home and fed scraps and maize are not a money-making venture. Indeed, my own rough calculations showed time and again that in the values of the marketplace, a home at best breaks even in its haphazard chicken raising. Chickens are, to use the phrase of one person, the 'savings account of us poor ones'. His terminology could not be more precise: unlike interest-bearing savings accounts in most banks, chickens are a true savings account, a store of value, a hoard, no more. And these animals of savings can, by a twist of the neck, be easily converted to consumption or 'rolled over' in value through the sale of mature ones and the raising of chicks. This is not to argue that chickens, hogs and cattle are not at times sources of profit, but their principal function within the use-value system is as storehouses of product. As more durable forms of value than agricultural produce, animals and other objects play an important part in helping to sustain a household through times of good and bad. Savings are kept in the production-consumption unit. They are not thrown into circulation.

Alternatively, a portion of the surfeit production may be used to make traditional purchases, such as consulting a curer. But these consumption desires are not ever-expanding; and whether the surfeit becomes savings or goods, neither hoards nor the purchase of non-basic commodities leads to growth: they are non-productive subtractions from the amounts produced.

The major portion of the surplus product is consumed and thus converted to time, the interval between simple reproduction and the remainder of the year. But this 'unused' time is not utilized to support yet more labour at productive work and to generate more surplus; the freed time is itself split up.

In the past, one small portion of this labour time was used to provide rental payment for the land. Even then, however, the rent - but a few days' labour - was small. A second portion of the released time is used to meet social duties: contributed labour to support the school and chapel or to cut the grass at the side of the highway. The first two activities are optional and may amount to two days a year, while the latter, known as a *faena*, is done at government order and may require but half a day per year. Some portion of the surplus time, third, may be consumed in the *fiesta* cycle. In addition to the fifty-two Sundays of the year, which are rest days, and Saturdays which are often half a day of work, there are approximately fifteen holidays which are invariably celebrated. Although the festal ceremonies vary, they always include a prohibition on men working in the fields. And beyond these days, there are at least twenty other *fiestas* which may or may not be kept at the volition of the individual. The religious calendar is flexible, and in case of pressing need a man will ignore the religious prescriptions, but it requires a not insignificant use of surplus time. Technically, at least a portion of these latter two forms of duties could be considered taxes and tithes, a form of surplus extraction by external agencies, though the impetus for performing them is mostly internally generated. All together, rent, social contributions and religious days account for approximately 100 days or over 25 per cent of the year.

A rather large amount of surplus time is still left, however, and this time is utilized in important ways within the countryside. A variable amount of time is used in other productive activities, such as fishing or making hats, fishnets or other objects for the home or sale. Other time may be spent on crucial subsistence work such as building or repairing a home, finding firewood, keeping up a house garden or securing needed consumer items.

Lastly, however, one block of time - 50 to 100 days a year - appears as true leisure, and this in two forms. Some days a man spends simply resting or relaxing; he may stay around the house or visit in the community. Interestingly all these activities are known as being '*de valde*' - around the tub. When a man is not working in the fields, he is by definition at home. Alternatively, leisure may be absorbed by spreading the necessary labour over a longer period through working more slowly or by stopping in the early afternoon. Indeed, when drawing a contrast between the traditional mode of production and a salary system the men often point out that they prefer the former, since it allows them to work on their own and not under the orders of someone else. It is, incidentally, this last block of 'leisure time' which presents the greatest potential resource for the capitalistic system; leisure time is pure surplus which may be utilized. From the peasants' standpoint any earnings gained in leisure time are an extra, an abundance, as they say. From the standpoint of capitalism what is required is a crop or industry which fits the traditional system such that its labour needs fall during the slack times of the agricultural cycle. As the peasant has already assured his subsistence, wages can be depressed without physically harming the labour force. The labour requirements of raising sugar cane, as it happens, fit well with the requirements for growing rice and maize for the home.

None the less, viewed only within the context of the subsistence system, it is the segment of leisure time that illustrates in sharpest form the non-accumulative nature of the economy. Of course, the technical requirements of agriculture do not always permit of steady work through the year, and thus the technology to some degree presents the opportunity for unused time. But the social relations of production also are not organized to make use of and to reinvest surplus labour or product.

Since production relations are internal to the household, increasing the rate or the absolute volume of surplus means increasing self-exploitation or the intensity of labour. But the increased product is of use initially only to the producer. There is no means by which to convert this household surplus to yet a further surplus within the household itself. Working for the self is not cumulative and is personally unpleasant. Within the household the only possibility for harnessing more labour lies in the father-son relationship, but it is precisely this relationship which bursts under the strain of surplus extraction. On the other hand, within the broader subsistence system household surpluses have little value in exchange, since others also are producing for consumption; and the kinship system itself is not constituted to appropriate surpluses. The self-sufficiency of production units assures that labour does not become a commodity and hence both a use for and a source of surplus outside the household.

The countryside, however, is not independent of the metropolitan area, and the lack of political and social institutions, the vacuum of traditional leadership, within the community is a function of the fact that the economy is embedded within, occupies a niche at the bottom of, the larger capitalistic society. Even the local 'alderman', always a community member, is appointed from the cities, and, perhaps more important, his authority is minimal precisely because he lacks an economic base; he is unpaid and has no power to tax. The social relations for converting surpluses to community projects or enhanced productivity have themselves been appropriated by the larger society. Cause and consequence, the self-sufficiency of the production units is linked to the paucity of an institutional superstructure, to the undeveloped extra-domestic relationships within the village.

Thus, we return once again to the distributional problem, the relation between subsistence and surplus. The rate of surplus remains relatively constant, just as in absolute terms there is little impulse to increase surplus product. These are economic facts. But a marginalist explanation of a subsistence level, such as 'the peasant works to the point where his last input of labour - its drudgery - is balanced by the reward of the final unit of product', is little more than an *ex post facto* restatement of the situation. The marginalist operates at the micro-level and what he leaves unexplained is why this particular balance is struck for the 'average' individual in this particular society. Against this line of argument I hold that an explanation of the subsistence level must be sought in its relation to the magnitude of the surplus and its use, and that an explanation of this broadscale distribution in turn must be sought within the overall patterning of the socio-cultural fabric and institutions of production. In fact, here the word understanding seems more appropriate than explanation. We cannot perhaps explain why a particular rate of surplus is found in a particular society; at most, we can attempt to understand the pattern by viewing it within an extra-economic context. Explanation enters only when a comparison can be drawn. Once a pattern of distribution is given in a society, what we can attempt to explain are the causes for its changes over time or the variations observed between it and the patterns of other societies - differences, it hardly need be added, which are the result mostly of socio-cultural factors.

Overall, then, the conditions of the Los Boquerones subsistence system continually re-create themselves and the production system remains relatively unchanging until the encompassing economy changes the provisions of its relationship with the peasants. In terms of the outside system their surplus labour awaits extraction.

4 The production process

The social organization of subsistence depends upon the production of certain consumable goods. The application of human labour to natural resources, itself determined in form by historical conditions, underlies the entire subsistence economy.

Three forms of productive activity may be distinguished. Basic subsistence agriculture, an expression of my choosing, refers to the work, undertaken by nearly everyone, of raising the central crops of rice and maize. The production of these two foodstuffs sustains the economy, and through the production of these crops, value and surplus, as well as the rate of surplus, are primarily determined. A large number of crops also are raised on a smaller scale, in the fields or at home. These auxiliary foodstuffs are seeded optionally, and though no one plants all possible crops every year, as a collection they make a fair contribution to the well-being of a household. Lastly, non-agricultural production, including support activities for agriculture and small-scale, craft-type work, is necessary for the functioning of the economy but secondary to the other forms as measured by allocated time. Not split up amongst specialists, all three productive forms are practised by the household group.

Basic subsistence agriculture

The term 'technology' I use to refer both to the material implements and to the productive activities of the agricultural cycle. The named tools designate existent and durable objects, while the labelled categories of production refer to physical actions or labour which occur at certain times in certain places. Besides facilitating communication, one function of the latter set of names is to serve as reminders and planning devices for the agricultural year.

The technology of the two-year agricultural cycle is traditional in that it probably has changed little over the past several hundred years. But tradition should not be equated with lack of choice, for the people face a diversity of farming decisions throughout the year. Tradition and custom are information, information used to set the available options and to help make selection from within these options. The distinction between the technology of subsistence and of capitalism ought not be reduced to the opposition 'simple' versus 'complex', for these terms, advanced from the vantage point of industrial life, pertain to the content of the technology. The difference rests more on the fact that one system uses a non-expanding information bank while the other thrives on the notion of progress, that the fund of productive ideas should expand, change, lead to greater productivity and yield greater profits. Of course, the technology of subsistence farming does change, and one can observe the effects of feedback from year to year, but the social organization of subsistence production does not itself provide an impulse for the change. New information about production is integrated on a haphazard basis rather than as a planned part of production. The system is geared to respond to, rather than seek, change. Certainly, the people are not slaves of their technological customs, any more or less than industrial man is a creature of his system, but their technology is not a conscious object of reflection and self-criticism. There is no reason for it to be.

Through the years the litany has been repeated that rural agriculture in Panama is stagnant and underproductive; an obvious remedy is to improve the technology of production. Indeed, at the grander level certain theories about agricultural development place special emphasis upon the flow of information to the peasant farmer (Rogers, 1969). But to emphasize the role of technology in change is to apply a framework derived from a totally different economy. The 'problem' of rural agriculture is not one of technology but of durable investment, more broadly of system difference. From the peasants' perspective their technology is well suited to subsistence, for it requires few cash expenditures on tools and no purchase of knowledge. To raise production in the interior through technological change would require a financial input, but this would mean that the portion of direct labour embodied in the product would fall, and the system would inevitably have to move toward a different criterion of success: return on investment. For the peasants the technological problem is not how to raise output, but how, under conditions of declining resources, to keep output constant, without a monetary investment, for such is the nature of their system and the virtue of their current technology.

Tools

The implements used are mechanically simple, being either hand-fashioned tools made from raw materials collected in the country-side or steel-tipped instruments purchased in the marketplace. Although the peasants have long been dependent upon an external supply of agricultural tools, implements made by profit-orientated concerns, these tools are durable and represent a small monetary outlay.

The simple nature of the implements is a reflection not only of the subsistence system alone but also of the relation between countryside and city; the rural area is dependent upon the city for equipment but the 'machinery' the latter provides, to be used effectively, requires a high labour input, which is the opposite aim of the factories which produce the equipment. The tools represent little accumulated or past labour, and no attempt is made by the peasants to refine and improve them, to make them more productive per labour input. Thus, in 1960, of the 21,231 farms in Veraguas, 3,736 were between half a hectare and two hectares in size. Considering these alone, only 2.9 per cent employed animal or mechanical power; human labour provided the energy source for the remainder (*Censos Nacionales de 1960*, 1965: 202).

The central and most important implement is the machete, found in different sizes and used for an endless variety of purposes. The largest, straight-edged type - called a 'colly' after the firm (Collins Steel) which makes them - is used for heavy tasks, such as cutting down the forest. A second form, with a broader but shorter blade, is termed a '*moruna*'.[1] *Morunas* come in various sizes, the larger ones being used for coarse agricultural work, the shorter for delicate weeding. Bent to facilitate the weeding, the short machete is also called a 'dagger'. But just as the full agricultural cycle consists of two years and work activities are distinguished by whether they are performed in the first or second year, tools also are called by different names after a year of usage, although their service years and the productive cycle may not coincide. In its second year the machete is known as *la mocha*, it has been 'maimed' or 'mutilated' after a year of work.

Fashioned from a long, straight pole, the digging stick is used to make holes for seeding. When fabricated solely of wood, the tool is known as a *chuzo*, while if a steel-tipped point is added it becomes a *coa*.

A water jug made from a local gourd is an essential to have when working in the fields. In addition, when weeding or cleaning scrub off the land, many use a *gancho*, which is a short, bent stick, stripped of its bark. Held in one hand, the *gancho* is used to pull back the brush and weeds to be cut by the machete which is held in the other hand. Another essential is the sharpening stone, usually found in the countryside or purchased from another rural dweller.

Maize is harvested by hand, while rice is collected by cutting the stalk with a small knife that is embedded in a wooden handle which itself is strapped to the hand. For transporting crops a basket, woven from vines and known as a *motete*, may be slung over the shoulders. But the materials and art of making *motetes* are fast being lost.

Finally, while the home contains many cooking implements, special mention should be made of the waist high, wooden mortar and three feet long wooden pestle; with this rural-made implement, the rice, in its last stage of production, is shelled for home consumption.

To question the 'adequacy' of the tools is a biased query in so far as the answer presupposes the use of costly equipment, but the thought may be entertained that other inexpensive tools might be used. A case in point is the hoe. Hoes are available and cost approximately the same as the other tools, but they are not a standard piece of equipment. Is this cultural selection due to 'ignorance' or to trial and rejection? My own inquiries support the latter answer. Certainly, the men state that using the hoe is easier on the back, and that, for tomatoes and lettuce, the hoe is a better instrument for weeding the earth. But in an area where the land is not soft, due to discontinuous use and soil conditions, and where roots and trunks are an inherent part of the field environment, the machete appears to be a 'superior' instrument of production. Being a shorter implement it is more 'labour intensive' in that the worker must use his own arm as a lever, but by the same token the machete permits finer control of the steel edge. Requiring perhaps a greater expenditure of human energy than the hoe, the machete permits the total work to be completed more quickly.

Similarly, it is not uncommon for peasants to own a team of oxen and use of the plough in Europe certainly predated the Conquest. Yet, utilization of the plough, even when driven by animal rather than mechanical power, requires the accumulation of dead labour, something which the relations of production are not organized to provide. Not only do oxen represent an investment, but ploughed land must be prepared and used on a yearly basis. I myself observed the obstacles which even a modern tractor faces when first used on the land; the barriers posed by initial land clearance and soil conditions are immense. In different terms, the 'pay-back period' for the use of the plough is long, and it is precisely such time periods, representing the accumulation of past labour, that the subsistence system is not constituted to support.

Of course, all this is not to argue that incremental improvements in the tools cannot be and are not at times made; it is to suggest, however, that the tools are a function of as well as a constraint upon the form of production. To focus on the physical tools and view them as a 'hindrance' to production in the countryside is to misunderstand their integral relation with the rest of the economy, for the implements are a result of the historical conditions in which the rural economy was created. They are as much a product of mercantile capitalism as they are a barrier to the expansion of industrial capitalism.

The agricultural cycle

Their timing set by the weather, the agricultural activities are not spread evenly over the calendar year. The Panamanian climate is characterized by a pronounced alternation of a wet and dry season. Beginning in April, the winter rains at first come lightly and sporadically. Later, they become a daily occurrence, usually in the afternoon. By November, however, the rains and humidity begin to taper off, and in December they cease entirely, giving way to the hot, dry summer which lasts until April. Winter foliage is lush and green, while in the summer the vegetation turns brown. But the weather also has an effect on the earth. Much of the Pacific watershed appears to be composed of a lateritic type soil. The land, subjected to baking and heavy rains, when used intensively and denuded of forest, erodes and forms a surface crust. In Los Boquerones small, lifeless gullies provide evidence of past weather conditions and soil overuse.

The agricultural cycle is divided into stages, each of which is labelled. The stages are not linked absolutely to specified calendrical dates, indeed decisions about timing are crucial, but there is a correspondence between time of year and type of work.

The cycle begins in December or January with a search for forest land; traditionally, a person knew in advance the general area in which he would work. The forest, known as *monte*, is land which has been allowed to rest for several years, ideally as long as ten. It should have thick trees and heavy scrub. The term '*monte*' is used not only for the actual forest but also the land area worked during the first year of the cycle. Thus, a man will say, 'I am going to the *monte*' when he is going to undertake any one of a number of agricultural tasks. The same word may be used for the shrubs which in the first year grow up in the crops and must be weeded.

In January and/or February the *monte* is cut down and allowed to dry. Known as *tumbando* (knocking down) or *socolando* (cutting down), the activity's timing depends in part on the thickness of the *monte* itself. If cut too early, more green shrubs may grow up in the field and have to be cut again. If cut too late, the wood may not have sufficient time to dry. A fulsome *monte* requires about six weeks of drying time. In the past, when cattle were more plentiful, they would be allowed to graze in the field after the cutting, at one stroke saving some labour and providing a food source for the animals. A 'street' or firebreak eventually is cut around the field. The width of one '*brazo*' - both arms extended - the firebreak is an area cleaned down to bare earth.

Although firing of the fields may begin in late February, the latter part of March is the height of the burning season. The firing must be done with some care, for in the hot summer there is the ever present danger that the flames will leap the breaks and spread to resting *monte*, crops or even houses. Every year, it seems, one or two fires grow out of control, leading in turn to further cutting of firebreaks or the loss of some resources.

Often fires are set at night, when it is cooler; always a man solicits the help of others. Sometimes the fire is set around the edges of the field in order to peak in the centre; other times a fire is set on the side farthest from the wind, and then reset on the other side to spread across the field. Since wind conditions, heat, thickness of the cut wood, length of drying time, and availability of help all determine when a field is burned, there is often a short lead time between deciding to burn and burning. But what sometimes appears to be a momentary, almost quixotic, decision, has been programmed far in advance; only the immediate timing is variable.

In the absence of capital equipment - in this case, large, earth-moving vehicles - the fire is a quick and inexpensive means for removing the forest cover. Burning the fields, however, is not seen by the people as an alternative to making a cash outlay, rather, what it saves for them is the physical labour which would be needed to remove the scrub. But burning is considered important for other reasons also. A general rule of thumb is that the bigger and stronger the fire, the better the land for raising crops. Burning, according to the people, gives strength to the earth, and the ash is a fertilizer; thus, the bigger the fire, the better the land. A good fire also tends to reduce the amount of weeds which grow up. The size of the fire itself is a function of the amount of dried wood, which in turn is a consequence of how long the land has rested; and, of course, the longer the land has rested, the more fertile the soil. Thus, the generalization, the ideology, of the people that the fire gives strength to the land actually is a condensed statement summarizing their ideas about fallowing, land fertility, and the amount of labour they anticipate will be needed to clear the weeds in the coming year.

Of course, reality does not always conform with technology. A grassland which yields a big fire should, by the rule of thumb, be good crop land, but everyone knows that it is not. Conversely, it is said that a very heavy fire may burn the topsoil, and thus a thick *monte* may give better results in its second year than in its first, contrary to normal expectations. And, when torn from context, bits and pieces of this traditional information system may facilitate the acceptance of 'modern' techniques. Although the peasants have been reluctant to use the herbicides and pesticides suggested by various outside agencies, they have been willing to purchase and spread fertilizer, since fertilizer and ash are equated in their ability to give strength to the land. Under conditions of forest depletion, fertilizer has become a natural substitute for fire. By contrast, herbicides and pesticides are not so easily assimilated to the traditional technology.

After the burning, the field is given a rather quick cleaning. Stones and unburned logs are thrown to one side, but large stumps are allowed to remain. Unburned shrubs may need to be cut or weeded.

From the middle of April to the middle of May is the time for seeding, and for this activity a small group is gathered. One person using a digging stick makes small holes throughout the entire planting area. The holes, about two to three inches deep and one and a half to two feet apart, are not made in rows. Carrying small calabashes of rice seed, the other workers drop into each hole several seeds and cover them over lightly by scratching the earth with a small stick. The seeding, it may be added, is the only time at which 'ritualistic' in addition to purely 'technical' actions are employed. As he begins his day's planting a man may utter briefly, 'In the name of the Virgin may it come up well.' A similar prayer is used upon arising in the morning or embarking on a journey.

The time immediately following the seeding is the most delicate period, for the rains are unpredictable. Too heavy a rain or prolonged dryness will kill the seeds. Birds and small animals also may ravage a field. Not infrequently a person has to replant at least a section of a field; occasionally in low land the year's entire crop is washed out.

Directly after the rice is seeded, maize and/or beans may be added to the field.

Weeding begins some three weeks to a month after seeding. By this time the rice sprouts have appeared and are clearly distinguishable from the other foliage which also is growing. The seeds have thrown off several stalks ('children') and the rice thus appears in small clumps throughout the entire field.

The general term for weeding is 'to clean', while the land itself is said to be 'dirty'. But the first weeding of rice is known in particular as '*desherbando*', literally weeding. Usually the first weeding does not begin until the ground is fairly wet, for the scrub is easier to cut then and there is some belief that hot dry earth falling on the rice hurts the crop. The task is backbreaking, for the worker proceeds through the field, bent at the waist. With a wrist and arm movement, he cuts the weeds with small strikes of a short machete. Each clump of rice must be cleaned all the way around. The weeds are left to dry where they fall.

The peasants recognize three degrees of this tiresome work. At times when a person is behind schedule, working for someone he does not like or is just lazy, he does little more than swing the dagger across the weeds and tramp them down. Known as *guachapiado* (from the verb meaning to splash water or bungle a task), this is the quickest and easiest way to weed. More time consuming but a better quality of work is *sobrepeine*, meaning to trim hair slightly. The worker cuts the weeds at their bases. Best of all is to cut the roots of the weeds an inch to an inch and a half under the ground. This is the slowest and hardest form of weeding and is called *casqueado*.[2]

Directly after the first weeding is completed or up to a month later, the second weeding is begun. This time the labour goes faster, for the land already has been worked, is softer, and there are fewer weeds. The second cleaning is labelled the '*repica*'. Only very rarely, for a long-maturing rice strain, is a third weeding given, and even then it is done quickly. By the end of July the year's weeding usually has been completed.

At the end of this same month the maize and beans may be ready for harvesting, if they were planted alone; the rice harvest begins in August and continues into early November, depending on the seed variety. Again, timing is important, for the rice must be harvested rather rapidly when it matures. Harvesting, called 'cutting' rice, is usually done by a group. Moving through the field, each person cuts the stalks about a foot below their grains. Gathering a handful of stalks, known as a *puño*, he places this aside, collects two more *puños* and then binds together all three to make a *manotada*, the traditional unit of counting and storage. After the rice has been 'cut', the maize which was interspersed with it is easily 'collected' (*coger*).

The harvested rice is transported to the house, left to dry outside for several days, and then placed in the loft above the kitchen fire. There, in *manotadas*, it continues to dry and can be taken as needed. Before using, the rice is dried again in the sun, then the grains are shaken from their stems and shelled. In *manotada* form rice lasts a year to a year and a half. Keeping the rice in the loft is both a form of storage and a method of drying, and for this last reason the rice is left on the stem until needed. By contrast, when the peasants do buy rice from storekeepers or millers, they pay not only for the product, the shelling, and the profit, but also for mechanical drying by hot air blowers, a process which costs them nothing in the countryside.

The bean harvests are rarely large enough to present a storage problem. Maize is kept on the cob in the leaf but will store for scarcely a year.

If, after the rice harvest, a second crop of maize or beans is to be planted, the land is cleared once again. Known as '*pelando*' (shearing off hair), the work consists of clearing the area to bare earth. Although physically similar to many other activities, *pelando* is said to be different in that it refers to clearing the land in times of wetness; hence, a fire cannot be used to reduce the labour required. The maize, it is said, must be seeded before 15 October since the rains cease shortly thereafter.

On the other hand, if the rice field contains a crop still growing, such as sugar cane, yuca or maize, but is not to be planted again, then the stalks of the rice are cut down (*despajando*), but the land is not cleaned so thoroughly as when it is to be planted again.

By the middle of November the work year begins to draw to a close. Often all that remains is to harvest the maize or another crop in January - but a few days' work; and the ripe maize, since it is now summer, is often left on the stalks to dry and be harvested intermittently.

The schedule for the second year is similar to that for the first, but with certain variations. The land is no longer known as *monte*, but as *rastrojo*, stubble. The term '*rastrojo*' is used for the rubble left in the field after the first year, for the plot throughout the entire second year, and for an abandoned plot which is regenerating. In the second year the land must be cleared again for burning, but now the work is called *apporeando rastrojo*, beating the stubble. In fact, the work is similar to the 'shearing', done in September and October of the first year, except that it is performed in the dry season and burning follows. A *rastrojo* usually produces a slightly smaller harvest but more weeds than a *monte*, in part, it is said, because the fire is smaller.

After the second year of use, the field is abandoned. It is still termed *rastrojo*, however, until the forest cover has returned sufficiently for it to be regarded as *monte*. Thus, aside from pasture land, house sites, and now sugar cane, all land is categorized as being *rastrojo* or *monte*. In themselves the terms have no absolute meaning, they are relative to one another. In fact, there is good evidence to suggest that their absolute meanings have changed over time. In the past, *monte* referred to land which had rested seven, eight or ten years. With the depletion of the forest, however, no land is left to regenerate this long and the term *monte* now is applied to fields which have rested but five or six years. In some areas of the interior yet a third term used to be employed, *tumba caña*. A field in its first years after use was known as a *rastrojo*. Then, it became *tumba caña* for three or four years until finally it grew to be *monte*. Thus, what people now call *monte* is what used to be termed *tumba caña*. With the shortening of the rest cycle, the category *tumba caña* has been squeezed out; owing to ecological pressures the linguistic label has been deleted. In the broadest sense, then, *monte* refers to land which is ready to be used in the agricultural cycle and to land which is being used in the first year. *Rastrojo* refers to land in the second year of the cycle and to land which cannot yet be 'reinvested'. As meaningful terms *monte* and *rastrojo* are related not only to real world objects but also to their positions in the productive process. Similarly, *desherbando* and *repicando*, as well as *socolando*, *pelando* and *apporeando* are alike as physical actions but distinguished by their places within the agricultural cycle. Keyed neither to specific dates nor physical objects, the meanings of the technological labels are tied to the immutable requirements of the productive process.

If one characteristic of the people's technology is its minimal incorporation of accumulated labour, the counterpoint is its susceptibility to the vagaries which afflict all farmers. Unlike corporate farmers, who may buy and sell grain futures to insure against potential failures, the peasants have few means of defence against total loss. In this regard the technology is not just a reflection of the production relations but a limit on them and on the production of surplus. The elements, of course, are always a problem. Too much rain or too little rain, at different times of the year, can have a severe effect on a crop. For high land much rain is desired, while just the reverse is true of low land. In June, July and November the wind can be a problem in that it may break the stalks of not yet mature maize. There appear to be few virulent diseases which afflict the crops, but birds and rodents, at different times and in different ways, attack every crop, from first seeding to harvest. Infrequently, mechanical means to ward off animals are employed, such as hanging rags in the fields or firing shots. Some home-made potions also are used, and the people buy marketed poisons. In the past, I was told, plagues, such as swarms of locusts, were not unknown. Some people would collect a few of the animals, take them to church to obtain a blessing, and then release the animals once again, thereby hoping to receive a supernatural solution. Overall, however, the technology leaves the peasant farmer rather defenceless against the unpredictable. But the people cope with this problem in part through their choices and decisions, for though the traditional technology is the same for all, there is much room for individual manoeuvre.

Alternatives and decisions

Throughout the year the peasant farmer faces a series of selections amongst which he must choose carefully if his livelihood is to be assured. Rather than say they are going 'to decide', a word which might sometimes be used to describe an agreement between two persons, the *campesinos* speak of 'thinking' about something. Thinking about and then doing something is an individual act which requires no arrangements with others.

The people do not reflect about the process of making choices in the same way that decision making has become, in some industrialized countries, an object of thought itself. Decisions are made after considerable thought but no ideology surrounds and glorifies the process of selection itself. This is not an economy which uses the image of marketplace man selecting among alternative resources.

Although information is not formally sought and evaluated, informal communications are extremely frequent. After work or in the evening men often socialize at one of the local stores or a house. The conversation varies from the telling of jokes and stories to gossip, but embedded within it, often a direct topic of discussion, is information about agriculture, the weather, crop problems, and what others are doing. The information system is closed in that the conversation is limited to topics which have been discussed year after year; however, within this sphere of discourse much relevant information is passed, in an almost haphazard manner. Decisions are private but the information on which to base them is public and free.

Directly at the beginning of the year, a major decision has to be made concerning the terrain to be worked. Since land quality varies over short distances, often a man considers not only how thick a *monte* is but how others have fared with nearby plots in the recent past.

smaller market price, sometimes as much as $3 per 100 pounds (one *quintal*). But *petaca* is raised as insurance within the subsistence system, just as a better rice is grown to eat in the home. By keeping the first-class rice and selling the *petaca*, the people are in fact 'maximizing their rewards', but in a system which is not profit directed. In the terminology of Adam Smith, the 'use value' of rice dominates its 'exchange value'.

Choosing a seed is only the beginning of a series of inter-related decisions. Depending upon the seed and wetness or dryness of the land, a man may decide to transplant. The seed bed may be at the house or in the fields. The major advantage of transplanting is that it allows finer control of the water source, since the seed bed, being small, can be sprinkled. Transplanting, like using different seeds, is a way of coping, through an extra labour input, with the uncertainties that arise at the boundaries of the technology.

Even more important are decisions about what other crops to plant. Both inter-cropping and double cropping are practised. Usually maize and often beans are added to a rice field, and for each of these crops there are various seed strains. But the permutations of crop combinations and sequences are nearly endless. A common variation is to seed a field with rice interspersed with maize. Then, after the harvests, the field is reseeded with a second crop, or as the people say, a second *coa*, of maize.[3] This option requires that a relatively fast-maturing rice seed be used, so that the field can be cleaned and seeded again before the rains cease. Alternatively, a corner of a field may be devoted to one crop alone, such as beans, or to a sequence of crops. Or, the field may be divided into two parts, alternating the crops in each half. Much, of course, depends on the land, and whether it is thought to be good for certain crops. In general, it is said to be difficult to reap sizeable harvests two years in a row practising both inter-cropping and double cropping, and of course the more crops that are added to a given area, the smaller the yield of any one crop. But usually the people try to pursue a programme of diversification, just as they do when selecting and combining varieties of a seed.

The agricultural year, thus, consists of a stream of inter-linked activities. For each person there is usually a logic, a greater or lesser degree of reasonableness, to what is undertaken and accomplished; but when the community is viewed as an entity, 'across the board', only a rough consistency among the different farmers makes itself manifest. On any single day very few are doing exactly the same task. For example, one September day, one man was preparing land near his house for planting maize, while another was giving his rice a final weeding; one individual was harvesting rice, and another was resting at home; one person was repairing some objects around his house, while a sixth was reseeding some sugar cane. Viewed thus, the systematic nature of the village economy is difficult to perceive. Indeed it exhibits a good deal of 'friction' in that activities between households are uncoordinated. Such 'inefficiency' of the larger scale seems endemic to a system of household production for use. Yet, the 'invisible hand' - the unplanned result of individual decisions - also yields one type of efficiency. Some of the agricultural tasks require the recruitment of a small labour force. If every economic unit were acting in parallel, none would be able to function, but since the work is inevitably, 'naturally', staggered among households, normally there are unoccupied farmers about who are willing to work for a day or two. Similarly, during the more intense times of the year an unoccupied man usually can find remunerated work, precisely because the community itself does not constitute an organized economic entity.

Profit, value and surplus: the functioning of the system

Most workers are able to provide their domestic units with the goods needed for the year, but household variation exists in the amount of labour expended and the volume of items produced in basic subsistence agriculture. We may here extend the discussion of the previous chapter by looking more exactly at the size of the agricultural surplus and at certain related quantities.

For these calculations I shall deploy two sets of figures: the 'socially expected' and the 'actual' (Tables 1-6). Collected from dozens of persons, there was a high degree of consensus about the socially expected data. These are the traditional standards or expectations that the peasants have about their basic subsistence work. Everyone 'knows', for example, that one hectare of land will require about 25 pounds of rice seed - 'the old ones told us the rate'. Based on accumulated experience, these work norms are in essence technological standards, the specifications of the work process. Comparable to the forecasts and budgets of a business firm, the rules of thumb are plans or guides. They tell what combination of inputs may be expected to yield a certain range of outputs. For each quantity I provide a range rather than a point calculation, not to show absolute extremes but to indicate the 'normal' variability which the people expect.

We may, however, understand the traditional extrapolations in a different sense. An abstraction from years of experience and hundreds of farmers, the cultural rules of thumb state 'the labour-time socially necessary' for production (Marx, 1967a: 39). The peasants may lack a turn-of-the-century Taylor (1911) to analyse their work procedures with the aim of making their efforts more efficient, but this has not impeded them from formulating cultural norms about their labour. Socially necessary labour time is not an abstraction of the outsider, it is a folk generalization.

The 'actual' data come from a different source. Of the ninety-one households in the community, fifteen were selected for study by a random process. A questionnaire, pre-tested and developed after a year of fieldwork, was administered by myself in early 1967 to each of the households. Usually, the 28-page form took two days to complete, not counting time for re-checking. Overall, I am confident about the 'quality' of this data, since I was often able to 'verify' the material through internal cross-checking, redundancy in the questionnaire, and external sources, and because the respondents had little difficulty providing the requested information.

For purposes here, certain assumptions about and simplifications to the raw figures have been made. First, both the actual and socially expected data refer only to the production of rice and maize. Since other crops are seeded with the same labour in the fields, though in varying and small amounts, this simplification is a 'conservative' one in that it leads to understating the value actually produced by each household. Second, the data cover only one year's usage of a field. In the case of the sample households, the fields were either in the first or second year of use. For the socially expected figures I have assumed the fields are being used in the first year (second-year costs would be lower in land preparation but higher in weeding, while harvests would drop slightly). Third, the household is taken as the unit of production, regardless of size. Labour costs include the efforts of all the workers recruited by the household; the number of household consumers appears, as I shall explain, in calculations about surplus. Fourth, all transfers between households have been omitted. Whether or not a portion of the crop was sold is irrelevant. How and in what form labour from outside the household was recruited and paid also is omitted. Fifth, measurements are given in labour days and product size. The former is the unit which the people themselves provide; the latter is calculated in pounds but may originally have been given to me in one of a number of traditional measurements which I have converted. Lastly, analysis of inter-household variation in the sample is not of concern here, but the reasons for such variation include relative size of domestic labour force and needs, the place of rice and maize in relation to other productive activities of the group, crop failure, and abilities of the household workers. The 'actual' figures for production include only fourteen households, for one man, though included in the overall sample, was no longer able to work at basic subsistence activities.

In Tables 1 and 2 are presented the essential costs and output figures for rice and maize. Though tiresome, a brief line-by-line explanation of some of the entries in these tables is required so that the figures may properly be understood and interpreted.

Rice, *Gross Product* designates the size of the threshed but unshelled and undried rice harvest. This is the total output for the year.

Rice, *Net Product*, by contrast, refers to the output after deduction of seed costs. This is the output after the initial stock has been reproduced.

Rice, *Net Product*, *Selling Price* is the dollar value which would be received if the net product were sold. In the sample, if the product or a portion of it was sold, the actual price received has been used to evaluate the market value of the total net product; otherwise, the net product is evaluated at $0.04 per pound, its usual selling price at harvest time.

Recently harvested and threshed rice (either gross or net) contains impurities, moisture and the weight of its shells. Before it can be consumed, it must be processed by hand or at a mill. Thus, only a portion of the net product is actual, edible rice and the net product figure must be adjusted to give its consumption equivalent. I have decided, after consulting with rice millers and the peasants, to use 0.6 as the conversion or reduction factor. The census bureau employs a slightly different and higher number - 0.66 - but their figure is applied after the rice has first been given an unknown amount of drying (*Estadistica Panameña*, 1967a:v). *Rice*, *Net*

Product, Consumption Equivalent, then, is what a household as a consumer unit receives from its harvest after replacement of stock and 'processing'.

TABLE 1 Socially expected: costs and yields of rice and maize for one hectare

Product output	Low yield	High yield
A. Rice		
Gross product	2,143 lbs	3,571 lbs
Less seed	−25	−30
Net product	2,118 lbs	3,541 lbs
Net product, selling price	$84.72	$141.64
Net product, consumption equivalent	1,270 lbs	2,125 lbs
Consumption equivalent buying price	$158.75	$265.62
B. Maize - first harvest		
Gross product	400 lbs	500 lbs
Less seed	−12	−20
Net product	388 lbs	480 lbs
Net product, selling price	$11.64	$14.40
Net product, buying price	$19.40	$24.00
C. Maize - second harvest		
Gross product	500 lbs	800 lbs
Less seed	−12	−25
Net product	488 lbs	775 lbs
Net product, selling price	$14.64	$23.25
Net product, buying price	$24.40	$38.75
Production costs in days		
Cut down forest	10	15
Cut firebreak	2	4
Clean after fire	5	3
Clean for second maize harvest	15	10
Seed rice	7	10
Seed first maize harvest	1.5	1.5
Seed second maize harvest	2	3
First weeding, rice and maize	20	10
Second weeding, rice and maize	12	8
Weeding, second maize harvest	8	6
Rice harvest	12	20
First maize harvest	2	3
Second maize harvest	2	4
Total Days	98.5	97.5

TABLE 2 Random sample: costs and yields of rice and maize

Household number	1	2	3	4	5	6
Product output						
A. Rice						
Gross product	4,564 lbs	651 lbs	500 lbs	225 lbs	8,094 lbs	675 lbs
Less seed	−50	−20	−12	−6	−63	−12
Net product	4,514 lbs	631 lbs	488 lbs	219 lbs	8,031 lbs	663 lbs
Net product, selling price	$187.80	$31.57	$19.52	$8.76	$321.00	$26.52
Net product, consumption equivalent	2,708 lbs	378.6 lbs	292.8 lbs	131.4 lbs	4,819 lbs	397.8 lbs
Net product, cons. equiv. buying price	$338.50	$47.32	$36.60	$16.42	$602.38	$49.72
B. Maize						
Gross product	610 lbs	816 lbs	66 lbs	66 lbs	1,600 lbs	200 lbs
Less seed	−14.5	−6	−2	−2	−35	−12
Net product	595.5 lbs	810 lbs	64 lbs	64 lbs	1,565 lbs	188 lbs
Net product, selling price	$19.34	$24.30	$1.92	$1.92	$46.95	$5.64
Net product, buying price	$29.78	$40.50	$3.20	$3.20	$78.25	$9.40
Production costs in days						
Total land preparation	45	20	5	4	55	21
Clean after fire	4	6	1	—	23	3
Total seeding	13.5	15	5	2	22	4
First weeding	18	12	6	3	15	8
Second weeding	5	15	4	2.5	3	9
Total harvesting (+ transport)	71	15	8	3	146	5
Total days	156.5	83	29	14.5	264	50

Rice, Net Product, Consumption Equivalent Buying Price is the total value of the net consumer product evaluated at $0.125 per pound, the fixed buying price for second-class rice. If the peasants were to buy their product on the market this is the amount they would have to pay. In that they usually raise a mixture of second- and first- ($0.15 per pound) class rice I again have been conservative in valuing their product.

Maize, Gross Product designates the total maize product off the cob.

Maize, Net Product refers to output after seed stock is replaced. This latter figure has also been evaluated at its selling ($0.03) and buying or consumption ($0.05) prices but no deduction need be made for processing.

Costs are expressed in labour days. In the sample 'land prepara-

7	8	9	10	11	12	13	14
3,250 lbs	750 lbs	430 lbs	1,300 lbs	3,193 lbs	2,500 lbs	3,333 lbs	5,365 lbs
−25	−30	−5	−25	−40	−25	−25	−75
3,225 lbs	720 lbs	425 lbs	1,275 lbs	3,153 lbs	2,475 lbs	3,308 lbs	5,290 lbs
$129.00	$28.80	$17.00	$51.00	$126.12	$99.00	$132.32	$211.60
1,935 lbs	432 lbs	255 lbs	765 lbs	1,892 lbs	1,485 lbs	1,985 lbs	3,174 lbs
$241.88	$54.00	$31.88	$95.62	$236.50	$185.62	$248.12	$396.75
800 lbs	40 lbs	1,000 lbs	200 lbs	400 lbs	1,000 lbs	300 lbs	700 lbs
−30	−15	−25	−6	−7	−40	−6	−18
770 lbs	25 lbs	975 lbs	194 lbs	393 lbs	960 lbs	294 lbs	682 lbs
$23.10	$.75	$29.25	$5.82	$11.79	$28.80	$8.82	$20.46
$38.50	$1.25	$48.75	$9.70	$19.65	$48.00	$14.70	$34.10
23	20	18	27	26	21	16	52
3	5	—	8	5	12	4	18
9.5	12.5	6	12	18	20	14.5	23
20	10	5	28	28	19	14	27
—	20	2	20	20	18	12	5
25	14	6.5	14	27	31	29	94
80.5	81.5	37.5	109	124	121	89.5	219

tion' includes cutting down the forest, making a firebreak and, when performed, cleaning the land for a second maize harvest. Seeding includes the time needed for both rice and maize.

One final cost, however, has not been included. In order to make the various calculations in product terms alone, I am excluding equipment costs, meaning 'the quantity of machete used up' in the productive cycle. In fact, this is a minor omission. Machetes cost about $1.25 and last two to three years, or on a 'straight-line' basis, machetes cost about $0.42 per year. Put differently, the yearly wear and tear of a machete is equivalent to about 3⅓ pounds of the net product consumption equivalent of rice, and that is a very small proportion of the product output.

Given this data we may turn to Tables 3, 4, and 5 in which are presented final calculations for the socially expected figures, the sample data, and a comparison of the two. Three categories of computations are provided: profit, value created per labour day, and surplus.

TABLE 3 Socially expected: profit, value and surplus

		Low yield	High yield
Book profit			
Selling price of total net product		$111.00	$179.29
Less labour cost @ $1.00/day		−98.50	−97.50
Selling profit		$12.50	$81.79
Buying price of product consumption equivalent		$202.55	$328.37
Less labour cost @ $1.00/day		−98.50	−97.50
Buying profit		$104.05	$230.87
Value created per labour day			
As product (net consumption equivalent ÷ number of days)	Rice	12.89 lbs	21.79 lbs
	Maize	8.89 lbs	12.87 lbs
As dollars [(net consumption equivalent rice × buying price + total net product maize × buying price) ÷ number of labour days]		$2.06	$3.37
Surplus			
Rice net product, consumption equivalent		1,270 lbs	2,125 lbs
Less necessary subsistence (3.6 lbs/day × number of days)		−355	−351
Rice surplus		915 lbs	1,774 lbs
Maize total net product		876 lbs	1,255 lbs
Less necessary subsistence (1.75 lbs/day × number of days)		−172	−171
Maize surplus		704 lbs	1,084 lbs
Rate of surplus = $\frac{\text{Surplus}}{\text{Necessary subsistence + seed}}$		2.69	4.75

$$= \frac{\text{Rice surplus} \times 1.66 + \text{maize surplus}}{(\text{Rice, subsistence + seed})\,(1.66) + (\text{Maize, subsistence + seed})}$$

Two different ways of calculating profit are given. The *Selling Profit* designates the difference between the sales value of the total (rice and maize) net product and labour costs (evaluated at $1.00 per day, the normal wage in rice or opportunity foregone). If the peasants actually were capitalistic farmers who sold their entire product and paid cash for hired labour, this is the profit they would have accrued. But this is a 'book profit', a fictitious figure, since no one sold his entire product nor hired all his labour. None the less the figure is not devoid of interest, for it may be compared with what I term the *Buying Profit*. Buying profit, another book figure, measures the difference between the market value of the net product consumption equivalent and labour costs, that is the difference between what it would cost the peasant to purchase his goods on the market and what it would cost him to produce these goods if he hired all his labour. In the context of home consumption, where an object's use value is more important than its exchange value, buying profit is a more accurate reflection than selling profit of the 'true value' of peasant production. Thus, whereas six of the fourteen households incurred a selling loss, only three had a buying loss. More striking, the average selling profit was $11.42 but the average buying profit was over nine times this amount. A comparison of these two profit figures provides some understanding as to why subsistence producers will continue to farm in situations where capitalistic firms will cease production. As the people point out, they make no profit on rice; they raise it because they must eat.[4]

The second set of calculations in Tables 3, 4 and 5 refers to the overall value which is created by a day of labour. This value is measured in quantities of use objects, pounds of rice and of maize. For later comparison to the sugar cane, however, I have converted these quantities to their market prices, or what it would cost to purchase them. On average, in the sample, each labour day created $1.86 in value, while the socially expected figures range between $2.06 and $3.37. And here again it is necessary to recall that my figures are conservative in that I have omitted the product value of the incidental seedings in a field.

The final calculations pertain to surplus and the rate of surplus. Again, in general terms surplus is defined as total product less the inputs of seed, labour and equipment. Specifically, given the preceding definitions and assumptions, surplus designates the difference between the *Net Product, Consumption Equivalent* and the subsistence necessary during the production period. For the sample I have 'standardized' the *Net Product, Consumption Equivalent* to a per day basis. From this daily product is subtracted the amount of rice or maize which each household actually consumes per day. (It is here, of course, that the number of consumers in a household has

TABLE 4 Random sample: profit, value and surplus

Household Number		1	2	3	4	5
Book profit						
Selling price of total net product		$207.14	$55.87	$21.44	$10.68	$367.95
Less labour cost @ $1.00/day		−156.50	−83.00	−29.00	−14.50	−264.00
Selling profit		$50.64	($27.13)	($7.56)	($3.82)	$103.95
Buying price of product consumption equivalent		$368.28	$87.82	$39.80	$19.62	$680.63
Less labour cost @ $1.00/day		−156.50	−83.00	−29.00	−14.50	−264.00
Buying profit		$211.78	$4.82	$10.80	$5.12	$416.63
Value created per labour day						
As product (net consumption equivalent ÷ number of days)	Rice	17.31 lbs	4.56 lbs	10.10 lbs	9.06 lbs	18.25 lbs
	Maize	3.80 lbs	9.75 lbs	2.21 lbs	4.41 lbs	5.92 lbs
As dollars [(net consumption equivalent rice × buying price + total net product maize × buying price) ÷ number of labour days]		$2.35	$1.06	$1.37	$1.35	$2.58
Surplus per labour day						
Rice produced per day		17.31 lbs	4.56 lbs	10.10 lbs	9.06 lbs	18.25 lbs
Less necessary subsistence per day		−4.00	−3.00	−5.00	−2.00	−4.00
Rice surplus per day		13.31 lbs	1.56 lbs	5.10 lbs	7.06 lbs	14.25 lbs
Maize produced per day		3.80 lbs	9.75 lbs	2.21 lbs	4.41 lbs	5.92 lbs
Less necessary subsistence per day		−2.00	−1.00	−2.00	−1.00	−4.00
Maize surplus per day		1.80 lbs	8.75 lbs	0.21 lbs	3.41 lbs	1.92 lbs
Rate of surplus = Surplus / (Necessary subsistence + seed)		2.63	1.76	.75	2.94	2.29

$$= \frac{(\text{Rice surplus/day})(1.66) + \text{maize surplus/day}}{(\text{Rice sub/day} + \text{rice seed/day})(1.66) + \text{maize sub/day} + \text{maize seed/day}}$$

a marked effect on the surplus.) For the sample the average surplus is 9.04 pounds of rice per day and 4.32 pounds of maize per day. For the socially expected figures I have again been conservative by assuming that 3.6 pounds of rice and 2 pounds of maize are consumed per day, higher consumption figures than in the sample.

If, however, we wish to add the rice and maize surpluses to derive the total surplus, or to calculate the overall rate of surplus, we again encounter the problem that objects of use are heterogeneous. How can an overall rate of surplus be calculated if pounds of rice must be added to pounds of maize? What transformation figure can be utilized to convert rice to maize, or maize to rice?

As neoclassicists we could value the rice and maize surpluses in

6	7	8	9	10	11	12	13	14
$32.16	$152.10	$29.55	$46.25	$56.82	$137.91	$127.80	$141.14	$232.06
−50.00	−80.50	−81.50	−37.50	−109.00	−124.00	−121.00	−89.50	−219.00
($17.84)	$71.60	($51.95)	$8.75	($52.18)	$13.91	$6.80	$51.64	$13.06
$59.12	$280.38	$55.25	$80.63	$105.32	$256.15	$233.62	$262.82	$430.85
−50.00	−80.50	−81.50	−37.50	−109.00	−124.00	−121.00	−89.50	−219.00
$9.12	$199.88	($26.25)	$43.13	($3.68)	$132.15	$112.62	$173.32	$211.85
7.96 lbs	24.04 lbs	5.30 lbs	6.8 lbs	7.02 lbs	15.26 lbs	12.27 lbs	22.18 lbs	14.49 lbs
3.76 lbs	9.56 lbs	.31 lbs	26 lbs	1.78 lbs	3.17 lbs	7.93 lbs	3.28 lbs	3.11 lbs
$1.18	$3.48	$0.68	$2.15	$0.97	$2.07	$1.93	$2.94	$1.97
7.96 lbs	24.04 lbs	5.30 lbs	6.80 lbs	7.02 lbs	15.26 lbs	12.27 lbs	22.18 lbs	14.49 lbs
−2.00	−2.00	−3.00	−5.00	−1.50	−5.00	−6.00	−3.50	−2.00
5.96 lbs	22.04 lbs	2.30 lbs	1.8 lbs	5.52 lbs	10.26 lbs	6.27 lbs	18.68 lbs	12.49 lbs
3.76 lbs	9.56 lbs	.31 lbs	26 lbs	1.78 lbs	3.17 lbs	7.93 lbs	3.28 lbs	3.11 lbs
−1.00	−1.00	−.50	−4	−0.00	−2.00	−3.00	−1.00	−2.00
2.76 lbs	8.56 lbs	(.19 lbs)	22 lbs	1.78 lbs	1.17 lbs	4.93 lbs	2.28 lbs	1.11 lbs
2.55	8.68	.58	1.89	3.73	1.67	1.12	4.54	3.66

terms of their market selling prices and then calculate the rate of surplus in homogeneous dollar terms. Similarly, since rice is purchased for $0.125 per pound and maize for $0.05 per pound, on the market 2.5 pounds of maize are equivalent to 1 pound of rice, or each rice unit is worth two and a half times each maize unit. But the problem with these solutions is that they allow the market to determine product value, whereas, surplus - the excess of output over input for time of production - is independent of market prices, as is the rate of surplus.

Alternatively, to define the conversion rate, we might use for each household the ratio of rice to maize which it produced per day. But here arises the problem that this ratio varies widely across the sample and differs each year. For example, in the sample, for household number one the per day production ratio of rice to maize is 4.56, but for household number two it is only 0.47. This suggests, in turn, that the socially expected production figures may provide a fairer, more stable approximation of the conversion ratio between rice and maize. Using these figures we find that 1 pound of rice is equivalent to between 1.45 and 1.69 pounds of maize.

TABLE 5 Socially expected and random sample compared: profit, value and surplus

	Sample average	Socially expected range		
		Low		High
Selling profit	$11.42	$12.50	–	$81.70
Buying profit	$105.93	$104.05	–	$230.87
Value created per labour day				
Rice	12.47 lbs	12.89 lbs	–	21.79 lbs
Maize	6.07 lbs	8.89 lbs	–	12.87 lbs
Value created per labour day - dollars	$1.86	$2.06	–	$3.37
Rice surplus per day	9.04 lbs	9.29 lbs	–	18.19 lbs
Maize surplus per day	4.32 lbs	7.15 lbs	–	11.12 lbs
Rate of surplus	2.77	2.69	–	4.75

This solution, however, raises another problem. The socially expected figures are based on the assumption that rice and maize are raised together in the same field; therefore, the relation of rice to maize output reflects the already existing valuation - the consumption desires - of the producers. What is required, by contrast, is a measure of the production relation between rice and maize which is independent of anticipated consumption.

To derive this conversion ratio the production processes of rice and maize must be treated separately. It must be assumed that one hectare is planted either in rice or in maize in order that the two purely production relations may be compared; we assume that the production process instead of yielding a joint product is devoted 100 per cent to rice or to maize. (In terms of the Sraffa system this is similar to the case when the whole of national income goes to wages and the values of commodities are in proportion to their labour cost (Sraffa, 1960: 12); however, unlike Sraffa I assume a non-uniform rate of profit (surplus) between rice and maize.)

This final transformation figure, based again on social expectations of the people, is calculated in Table 6. By these data each unit of labour yields one unit of rice or 1.66 units of maize. This value ratio, which is based solely on production and which I shall use to convert the one basic commodity to the other, is lower than the market ratio but higher than the socially expected ratios based on a mixed plot.

TABLE 6 Conversion ratio of rice and maize based on production for one hectare

	Rice	Maize
Product output		
Gross product	2,500 lbs	1,000 lbs
Less seed	−25	−30
Net product	2,275 lbs	970 lbs
Net product, consumption equivalent	1,485 lbs	970 lbs
Production costs in days		
Cut down forest	10	10
Cut firebreak	2	2
Clean after fire	5	—
Seeding	7	4
First weeding	20	8
Second weeding	12	—
Harvest	14	3.5
Total Days	70	27.5

Therefore: $\frac{1,485}{70}$ Rice $=$ $\frac{970}{27.5}$ Maize

1 Rice = 1.66 Maize
.6 Rice = 1 Maize

We may, then, return to Tables 3, 4, and 5 for consideration of the rate of surplus. This rate, it will be recalled, refers to the surplus divided by the necessary subsistence plus seed and equipment costs. (Again, the final category is omitted in the actual figures.) I have also suggested that this ratio corresponds roughly to a rate of profit, when no 'capital' equipment lasting more than a year is utilized. But this ratio refers to something more profound than a rate of profit. It is a measure of the return on investment and depends on both productivity and the apportionment of output between subsistence and surplus. The rate of surplus is a comparative measure, independent of prices, which summarizes a household's production in terms of its needs and capacities. The sample average, 2.77, characterizes the economy. By way of a broader comparison, Marx often expressed indignation at a rate of exploitation (surplus divided by cost of labour power) of 1.00, while a modern business firm often has an investment return of but 0.10 on total capital advanced. To conclude from these comparisons, however, that life in Los Boquerones is 'affluent' (Sahlins, 1972) would be misleading.

Affluence depends not only upon the relative rate of return but on its absolute level and the work needed to achieve this. In this regard, one final set of comparisons underlines the effects of the recent, sad changes in Los Boquerones and the decline of local 'affluence'. The socially expected standards, based on traditional experience, afford a glimpse at the nature and magnitude of the transition, for the traditional norms may be compared with the actual amounts of labour and product which currently characterize the system in its declining years. To be sure, all such comparisons must be undertaken with caution - to avoid the 'golden age, things were better then' phenomenon - but to dismiss the socially expected standards would be a greater mistake, given the consensus and recent experience which underlie them.

With the diminution of available forest, the people were faced with an alternative in subsistence production. On the one hand, they could have reduced the amount of land worked in a year to allow the use-rest-use cycle of the forest to remain in balance. But this response would have led to smaller yearly harvests, and, as can be seen from the socially expected figures of Table 1, under the best of conditions half a hectare will not normally yield enough food to last the year. Consequently, the people chose the other option of keeping plot size constant (or increasing it slightly) while speeding up the cycle; thus, the rest period has been shortened from 10 to 8 to 6 to 5 years.

Speeding up the cycle, however, lowers the land fertility, and this change has brought in its wake two further consequences. Harvest sizes have decreased both absolutely and in relation to the amount of labour invested; to produce the same size harvest as formerly, the people must work more. Their pithy summary of the situation conceals two different effects: 'We work harder now' - output proportionately increases more slowly than the labour input or output declines at the margin - 'and harvest less' - over time comparable labour inputs produce smaller outputs. The lament is borne out by a comparison of the average and socially expected figures in Table 5. For each measure the sample averages lie either outside or at the bottom end of the social expectations. Just as increased productivity decreases the amount of labour contained in an item, so resource depletion increases the labour content in a unit of agricultural product. Formerly it took between 2.0 and 3.3 per cent of a working day to produce one maize equivalent pound. Now, it takes 3.7 per cent of the day. The amount of goods created by the labour day has declined; the labour embodied in a unit of product has increased. Surplus has diminished and as one worker said, 'there is less free time now'.

In addition to its impact on labour input and product output, resource depletion has had an effect on the product mix. The land will no longer support a broad range of seedlings; fewer incidental crops, especially beans, are now raised, and of those crops seeded their harvests have also declined. Certain options of using the land, such as seeding beans and then rice, cannot even be practised. But the change in product mix can be seen most strikingly in terms of maize. Previously the expectation was that everyone would raise two crops of maize in the year; in 1966, however, only six households in the sample were able to do so. In order to preserve the absolute essential - rice - maize production has had to be curtailed. Thus, in Table 5 it may be seen that the rice created by a labour day falls just below the socially expected range, but the maize produced is nearly 32 per cent less than the minimum of the social expectations. Similarly, the average of the sample maize surpluses per day falls far lower than the range of the socially expected, but the same is not true of rice. Understandably, maize is no longer used as fodder for pigs and horses.

Thus, the subsistence system is reaching a perilous state, not only because of the decline in harvest sizes and increased work, but also because the people are becoming mono-crop dependent within subsistence itself. None the less, the production of auxiliary crops has not yet ceased.

Auxiliary subsistence agriculture

Auxiliary crops, which provide intermittent nourishment and some insurance, are raised on a casual basis. The harvests are used in the home, but individual crop failure does not lead to a household crisis, while a surfeit may remain unconverted and thus be lost.

The auxiliary crops include both annuals, which must be re-seeded each year, and perennials, which yield a harvest one or more times a year for several years. Depending on the crop, it is planted either in the fields alongside or interspersed with the basic ones, or in the garden around the house. The local ideology is that some crops need the 'heat of people', meaning sporadic but continual care, fertilization by household rubbish, and occasional weeding. These crops must be seeded at home. Fruit trees in particular are usually planted in the garden.

The principal classes of non-basic crops are tubers, fruits, and spices and herbs. In the sample survey, including all 15 households, the most crops raised by any one household was 47, while the least was 12. Nine of the households seeded more than 30 different crops, and the average per household was 28. The crops rice, maize, yuca and avocados were seeded by 14 households, while achiote (a food colouring), marañones, and mangoes were produced by at least 13 homes. At the other end of the scale a type of sweet potato and a bean variety were found only once, while in the middle, yams as well as the guanábana fruit (of the Annona family) were raised by 7 households. In addition, nearly all households seed one or another plant which can be used for medicinal purposes, the average being 5.5, the most being 14. Excluding medicines, the total number of different crops raised by all households in the sample was 50.

Except for the sugar cane all the crops are grown for the house and not for the market. Sales are considered incidental or secondary to home consumption. There is a seeming paradox here in that the peasants, desirous of cash, could sell some of their auxiliary crops, but they make little attempt to do so, except on a haphazard basis. Understandably rice, maize and beans are kept for the domestic unit, but why is not a surfeit of yuca or avocados automatically sold?

The reason why a particular crop is not sold dissolves into a series of factors, which vary for each seeding. One of the most important is small production volume. Some of the crops do not grow well on the land available, are frequently attacked by pests, require soft land, or need irrigation. For example, small tomatoes are raised by twelve households, but to grow them on a commercial scale would require far greater inputs of both labour and money. Compounding the volume problem is the fact that many harvests are intermittent. From a use standpoint this is advantageous in that storage does not become a problem, but the small quantities are difficult to market, and with a lack of storage facilities harvests cannot be agglomerated. Selling costs further hinder the marketing of local products. Crops to be sold either are taken to the highway and left until a passing driver stops and blows his vehicle's horn, or are carried in person by commercial transport to Santiago. In the former case, someone must be available to bargain and sell the crop to the prospective buyer. In the latter, the transportation raises the cost of the item and makes volume sales difficult to achieve, since it is hard to carry a large quantity on the small passenger buses. Then, in Santiago it may take hours to find a buyer for the produce; and the buyer, knowing that the peasant does not want to carry the item back home, indeed is selling because he desires immediate cash, has the upper hand in bargaining. As commodity sellers, the peasants have little withholding power. Thus, either method of selling requires time, a labour input which further reduces the 'profit' received.

In addition, market demand is often poor. For example, demand is non-existent for the marañon, a delicate fruit as well as cashew nut, which is used in various ways in the countryside. Small limes, which are rather plentiful and often made into a drink by the peasants, have only a limited market. The demand for coconuts is sporadic, since they are used primarily in certain festal foods.

Lastly, market prices in general are not high enough to draw foods out of the rural areas. The people themselves point out that the money received for a product usually buys less food value than that contained in the item sold. For them there is no incremental advantage in selling.

The growing and selling of yuca (sweet manioc) provides a good example of the problems encountered when selling a subsistence crop on the market. Yuca is seeded in the fields, interspersed with rice and maize, or in a small patch by itself. In May a man often seeds as many as 200 to 500 plants to be harvested in the following summer. If left in the ground yuca does not spoil, but when brought to the house it must be consumed within twenty-four hours. Used in soups and stews, or cooked and eaten alone, one of the advantages of yuca is that it provides the peasants with a ready food source in times of shortage. When placed on the market yuca is sold by the sackful, and a sack may contain forty to fifty tubers, a sizeable portion of the crop. Gathering the harvest, loading the sack and bringing it home take several hours; then, selling the crop on the highway requires more time. The final selling price varies but is usually about $0.75 per sack, roughly the equivalent of three-quarters of a day's labour. Thus, the people say that it does not pay to sell yuca. Only if a man has no food at all - has true necessity - will he sell yuca. Yuca, therefore, remains a subsistence crop not for moral or ideological reasons; the peasants are not averse to earning cash nor are they incapable of making monetary calculations. But since yuca yields a poor price and is difficult to market, it is planted primarily to supplement the domestic diet and diversify risk.

More broadly, value may be created only within the production process, but it can be extracted in exchange, when prices are depressed below true cost or labour embodied. Traders, located at the juncture of a subsistence and capitalistic system, can indeed reap profits through the purchase and sale of commodities. The disjuncture, the vacuum, which exists between auxiliary subsistence production and the commercial market is the result of numerous problems including inadequate transportation and storage facilities, sparse market demand, and low prices. But even a concerted attack on these empirical problems - it was once proposed to build a series of storage and market sheds throughout the interior - is unlikely to change the situation, for ultimately the subsistence system is not poised to produce food for the urban areas, although some always seeps out. Indeed, large-scale selling of extra foodstuffs is itself a contradiction. The people make no advance arrangements to market food, selling only when they have enough for the house and are in need of cash. But having made no prior arrangements, when the peasants do sell they are faced by the very problems which hinder sales. They often lose a saleable surfeit precisely because they do not normally sell. Only when the outside system invests cash directly in peasant agriculture and penetrates into the production process itself do the farmers consistently sell on the national market, as has happened with the sugar cane. Otherwise, the market has little to reap from peasant production except trader's profits - a fact perceived by the peasants and thus inhibiting sales, except in times of need.

Other productive activities

To survive the peasant producer must have the ability to undertake diverse tasks supernumerary to agriculture. The specializations of the countrymen, which include providing services and products, may be divided into three spheres: the domestic, the rural and the market directed. The first sphere refers to production for the home whereas the other two are for exchange: the first two spheres are confined to the rural area, while the third links countryside with the towns. Few of the specializations require costly equipment. In Table 7 is provided a list of the specialities practised by persons in the community.

TABLE 7 Non-agricultural occupations

1	Making saddle blankets of reeds for market sale
2	Making wooden plates, mortars and pestles, etc. for use and rural sale
3	Making thatch and mud houses for use and rural sale
4	Making bird cages for use
5	Making leather sandals for use
6	Making carrying baskets for rural sale
7	Making fishnets for rural sale
8	Weaving hats for use and rural sale
9	Making thread and rope for use and rural sale
10	Making small carrying pouches for use
11	Making buttons for use
12	Making wooden benches and other furniture for rural sale
13	Making and painting drinking gourds and other vessels for use and rural sale
14	Making and sewing clothes for use and rural sale
15	Making and selling food for rural *fiestas*
16	Making and selling *chicha* (corn liquor)
17	Owning and operating a horse driven sugar-cane press for product sale
18	Providing haircuts for cash
19	Cutting and hauling firewood for cash
20	Washing clothes for others for cash
21	Owning and operating a small rural store to make a profit
22	Owning and using a small, clay oven to bake bread for rural sale
23	Keeping chickens in a cage for sale on the market
24	Keeping hogs in a pen for sale on the market
25	Owning and operating a truck for hauling sugar cane for cash
26	Driving a truck or other vehicle for cash

Table 7 Cont.

27	Working (at times) in a restaurant, service station or on a highway crew for cash
28	Running a private lottery
29	Operating rural card games for cash
30	Working in the pasture of another for cash
31	Buying rice in the countryside and selling it to the mills
32	Buying and selling various items in the countryside
33	Slaughtering and selling meat in the countryside
34	Reciting prayers at funerals for cash
35	Curing by use of orations or medicines for cash
36	Fishing for home consumption
37	Hunting for home consumption
38	Shelling rice for others for cash
39	Working as a peon for others for cash

Specializations about the home, performed neither for exchange nor remuneration, usually are divided by the fundamental separation between the sexes. In addition to the domestic duties of raising children, cooking, washing, bringing water and caring for the house, a woman has other tasks. She may wholly or in part care for the household animals. Some women can make clothes, and a few make hats in their spare time. Food preparation itself includes more than cooking, for a woman must perform the final stages of food processing, by shelling and cleaning rice, a one- to two-hour task each day. A male similarly must be a handyman and perform diverse tasks just as he plants a variety of crops. A thatch house needs constant upkeep. In their spare time some men fish with pole or net, and a few still hunt small animals and birds.

This same adaptability of the human, this capacity to be a handyman, is also imputed to inanimate things, to be 'handy-objects'. In particular, purchased items, sometimes the castaway objects of industrial society, are adapted to new uses in the countryside. Old oil cans are fabricated into kerosene lamps; large soya bean oil cans make excellent vessels for carrying water and rice. Most remarkable are the uses to which corrugated tin is put: besides its use as roofing, it is shaped into gutters or walls, used to dry rice on, or made into temporary storage shelters. The observation may be banal, but it is important: the successful peasant, the coping peasant, can undertake an immense range of life-sustaining activities. Dependence on a specialization marks the end of the peasant role.

Traditional to the countryside also are certain specializations which are undertaken to earn cash, specializations which are seldom total occupations but which are important to the functioning of the overall rural economy. Women sometimes perform domestic duties for others: bachelors may pay a small amount of food or cash to have their food prepared; or if a female is busy, she may pay another from $0.10 to $0.50 a week to shell rice or wash clothes for her. Some women are midwives (although their task has been almost completely supplanted by hospitals), others sell lottery tickets or make clothes or *chicha* (corn liquor) for sale. Such tasks are not undertaken on a regular basis, but they do supply a house with supplementary income. Amongst the men some hire themselves out for $1.50 to $2.00 a day to make a house or crude wooden furniture. A few men know how to recite certain prayers and are employed during wakes, for about $0.50 a night, to lead the congregation in prayers and responses. Every once in a while a male will undertake a more capitalistic venture; one man, for example, several times made intricate arrangements to obtain a hog, slaughter it and sell the meat and fat. He usually found it difficult to sell all the product, and on a total revenue of slightly less than $80, he would make for his day or two of labour $1.00 to $2.00 plus some meat.

Three men in the community possess small, horse-driven, sugar-cane presses and cooking vats. Total investment for each set, which usually is fifty to sixty years old, amounts to about $300. The traditional product is a thick molasses or a brown sugar cake. One man, working three days a week, has yearly sales of $500–$600. But since this is his gross sales before all costs - equipment, upkeep, maintenance of the horse, growing and cutting the sugar cane, transportation to small stores in the countryside, and his own time for milling, cooking and selling - his remuneration on average is about $2.00 per working day. (Increasingly, the peasants are purchasing on the market white refined sugar in place of the traditional sweetener.).

The only other traditional occupation which requires a substantial investment is owning and driving an ox team and cart, the equipment usually being purchased now through a government loan. Driving an oxcart is not a full-time task, but teams are used to haul wood, harvests, sugar-cane seed, and construction materials for houses.

Certain of these occupations require learning but the skill is not automatically passed down in a family. Usually a brief apprenticeship with a friend or kinsman suffices for a person to become a skilled artisan. Many males know how to drive oxen, refine sugar, slaughter animals, recite prayers, or bake bread even if they have not practised the activity in the last ten years. The difficulty of performing certain tasks should not be underestimated, but there never has been a skilled elite in the countryside, a fundamental separation of the people based on differential access to special labour skills.

Lastly, some of the rural jobs are orientated towards the external sector. In general, these tend to be less traditional tasks, but they usually are not of such importance that they undermine basic agricultural production. One class of these market-focused jobs includes short-term ventures which hit the interstices, the temporary labour-intensive niches, in the national market. For example, one man occasionally would collect wild orchids, carry them to Panama City and hawk them in the streets. Another, under the supervision of a government agent, undertook to raise for sale a batch of caged chickens; but the technical problems - obtaining medicines and feed - and the costs, which seemed always to mount, overwhelmed him within a few weeks. Finding no ready market for their avocados two men, one of whom owns a truck, gathered a large batch of avocados and mangoes, transported them to Panama City, rented a shed in the marketplace and sold their produce. Their venture was successful, but it was based on an uncertain supply and if done on a permanent basis would have removed them from being farmers, from raising their essential subsistence goods.

When one's seedings fail, a popular venture from June to October is making reed mats for use under saddles. In the lagoon area of the community reeds may be secured for free. A man will spend several hours cutting and drying the reeds. After carrying several loads back to the house, he ties bunches of the rush together, using vines or inexpensive cord, to make a fairly durable horse mat. By working all day, often late into the evening, a person can finish as many as 100 mats a week, each of which sells for $0.15 to $0.25. The completed product is carried to the highway where it is purchased by transport drivers who resell it elsewhere. The job, however, is dependent upon the seasonal reed supply, and by 1967 intensive use began to diminish the yearly supply of the raw material.

A second class of external jobs is focused about the juncture of the rural and urban sectors. A few men own and operate ten-ton trucks for hauling sugar cane to the mills. During the harvesting season this is a full-time occupation. There are a few small stores in the village which sell market-produced items that are in highest demand among the rural populace. Located in the home, and owned by men, the stores are tended by women. Lastly, one man, owing to a kin connection, is a rice factor for one of the rice mills in Santiago. During the harvesting season he offers the going buying price - which varies by the week - and sells to the mill for a higher amount. He purchases in Los Boquerones and an adjacent community. In a poor harvest year his volume is about 300 *quintales*, rising to 500 in a good year, which yields him a cash profit of $75 to slightly over $100 per year.

The final class of outside orientated jobs is wage-labour, a form of work which has been expanding in the past five to ten years. Some of the men have been able to obtain occasional work with highway crews. One young man has been a service station attendant in Santiago for a few years, while others have become more and more dependent upon driving trucks or local buses. But, excluding work in the sugar cane, the job opportunities are few and have arisen primarily as a result of the completion of the Inter-American Highway. They are linked to the expanding transportation network.

The production of goods or services for exchange, whether with other peasants or urbanites, tends to be an impermanent occupation. But this is not a function of low motivation or some supposed inability to persevere. (Many of these tasks actually are tiresome, repetitive, and must be performed for ten to twelve hours a day.) Rather, lacking investment funds, the peasant can offer little more than his labour, and the countryside itself, being fragmented into thousands of independent units, has no traditional organizations which might be adapted to modern functions. Small production units which diversify their tasks among many labour-intensive activities are traditional to the rural area. These characteristics, which are so well suited and fundamental to a subsistence economy, are not in synchrony with the organizational requirements of a capitalistic economy. The forces of the market, with their increasing albeit piecemeal effects, will certainly change but not renovate the subsistence structure of the countryside. Thus far, in terms of production, the national market has drawn only upon unskilled rural labour, and even this labour has been utilized on an impermanent basis.[5] The national economy has not fostered the development of potentially competitive peasant organisations. Fragmented into domestic groups which are spread across the land, the countrymen can participate in the national economy only by working sporadically for others.

5 Organizing a labour force

In subsistence agriculture if the essential natural resources, the seed and forest, are in sufficient supply, the determining production factor is the amount of labour invested. During the agricultural year a household, the production unit, usually supplies and receives labour help from other such groups in the community. The volume and duration as well as the reward paid for this aid vary, but the ways of mobilizing and organizing a work force are given by custom, by tradition. The transfer of labour between households is a patterned transaction.

Upon examining forms of labour aid, we mark a shift, a 'surfacing', from the sphere of production to that of exchange. But the route between the two resembles more a labyrinth than a straight line, for there exists, first, the possibility that the modes of organizing community labour represent not simply inter-household transfers but units of production more important than the household, institutions more basic to the economy than the domestic units themselves.

On the other hand, if these communal labour forms are 'household based', are simply means by which households fulfil their necessary tasks, do they contradict the voiced goal of household self-sufficence? The unequal transfer of labour between households, the permanent extraction of labour power between the constituent units, could lead to differential accumulation of value, and this transformation would itself quickly undermine the postulates of household equality and independence, in short, the entire subsistence system.

The modes of organizing labour in fact do not emanate from institutions other than the household, nor are they simply an ideological justification masking the ugly reality of extraction. But the relation of labour exchange to the productive process is not a simple one. In part, the ways of mobilizing labour manifest subsistence alone; but subsistence is not without its contradictions, produced by differential, albeit impermanent, access to natural resources, by inequalities in the holdings of wealth objects, and by the tendency of the independent production units to disperse socially if not physically. These contrary forces within subsistence are projected into and actually cushioned by the patterned ways of exchanging and transferring labour between households. But the subsistence mode, generated by and encompassed within the market system, has always contained elements of that other economy. The capitalistic elements intrude not so much at the base level of the production unit, the household, as at the layer of exchange, the mobilization of extra-domestic labour. The market system is kept, as it were, a step removed from contradicting 'pure subsistence'. Lastly, in the recent past as the countryside has been subjected increasingly to the pressures of capitalism, the traditional forms of organizing labour have been changing. Their frequency of utilization has shifted, while as concepts they no longer match production practices. Paradoxically, however, the current transformation from subsistence toward the market reveals, by stripping away the veneer of labour exchange, that the fundamental production units have, in fact, always been domestic groups. Thus, the current labour patterns represent a blending of three modes: subsistence, subsistence within capitalism, and the change from subsistence to capitalism.

Production requirements

Ecological conditions and the agricultural cycle, that is the production function, plus the constitution of the production units and physical limitations of the individual conduce to the formation of labour groups, although they do not determine their form.

Household productive activities, as seen, are divided by age and sex. Since crops are primarily the responsibility of males, females helping only in the seeding and harvesting of rice, men must rely primarily upon other males for agricultural aid. Although women play a crucial role in the functioning of the overall economy, most forms of labour organization concern men.

Amongst productive males in agriculture, however, there is no specialization by job and no recruiting by speciality. Some, of course, have work preferences, some perform better at certain tasks than others, and in the course of a day's work there may be a minimal though impermanent division of labour, but the modes of labour organization are not based upon the inter-relation of specialists.

Leaving to one side the cash crop, which is worked almost entirely by wage labour paid for by mill loans, and given the constitution of the economy itself, the volume of household labour invested in the year falls within a range set by the number of domestic consumers and workers (Chayanov, 1966). At the minimum end, sufficient rice (and maize) must be raised to last the year. Although the ecological conditions are changing and the land is less fertile, the traditional norm has been that one hectare of fully seeded land yields enough rice and maize to last a four- to five-person family one year. Precise measurements of the land are not taken, but seldom do the people misjudge the minimum land area they will need to seed. Plot size, in turn, determines the minimum amount of labour that will need to be invested.

The number of household workers and lack of compulsion to produce a vendible surfeit, on the other hand, set the boundary on the maximum amount of land which is worked. Unless a man has several grown sons living in his house, he rarely seeds more than one and a half hectares in subsistence crops, and even this amount would be considered large, unless family needs made it necessary. Since working sons invariably leave a house to form their own production units, augmentation of crop land through the use of domestic labour alone is not a real possibility. To seed a very large area, then, a household head would have to invest in hired labour; but few have the cash resources to do so, and even fewer wish to risk their money in what is perceived to be an unnecessary venture. If a household head does have surplus time which he wants to use, he usually prefers to undertake a productive activity around the home or to hire out his own labour for cash which will yield money for purchasing household goods. Thus, subsistence activities are delimited primarily by household demand, by the quantity of labour which each household contains, and by the lack of compulsion to produce an agricultural surfeit.

Given these overall constraints on all domestic units, and within this range there are indeed household differences, we may consider in greater detail the required amount of labour which is traditionally expected for each agricultural activity. The number of agricultural options is large, but the traditional sequence was and is to seed one crop of rice and two of maize. In Table 1, to which I have already referred, is listed the range, the minimum and maximum number, of man days required to prepare, seed and harvest one hectare of first-year land in rice and maize. An examination of the table provides an overall view of the critical, bottleneck, farming tasks.

Usually, there is sufficient time for a man to chop down the forest by himself, but often he recruits others to help, since the work is hard and may require a joint effort. For the burning a man solicits the help of two to four others, but only for a few hours; cleaning the land afterwards can be done alone. Seeding does not require a large overall amount of labour, but the task must be done within a relatively short time interval; hence, the normal practice is to recruit a labour force large enough so that the field can be seeded in one or two days. The weeding also cannot be postponed nor stretched over too long a time, and a field badly invaded by weeds will be tackled by more than one man. Both the first rice harvest and the first harvest of maize must be gathered promptly if damage by rain or pests is to be avoided, and a labour force is also mobilized for these tasks. Most of the work in the final stages of the cycle can be accomplished by one man working alone.

Thus, labour groups are mobilized usually to complete tasks which are too heavy or too large for the single individual to undertake or which must be finished quickly. Since the land qualities, option selected and timing vary by household, work activities are staggered among domestic units, permitting the recruitment of labour. From the communal standpoint labour mobilization smooths out work capacities between households.

Beyond these requirements set by the cycle, the land, the capacity of the individual and the household division of labour, men express the desire to work together for conviviality and because 'it makes the work go faster'. Alone, a labourer looks ahead and thinks with dismay about how much work remains to be finished, but in a group he is diverted. Few people claim, however, that a group accomplishes more than the sum of its separate workers or than one man working alone for an equal number of days. Indeed, by careful planning and luck, some men do accomplish many of the necessary tasks in solitude.

Units of labour measurement

The fact that labour is the traditional measurement of value in the countryside is suggested by the two ways in which work itself is measured: by the labour job or task (*por tarea*) and by the labour day (*por día*). These quantification units cross-cut the ways by which work groups are organized; a group of labourers may work by the day, by the job, or by both.

Task work is piece-rate payment, being a specified job to be completed by one person before compensation is made. Ordinarily, the task can be finished within a day. The actual unit differs by the work and conditions encountered, and usually is calculated as a land area or an amount to be produced. For example, the generally accepted task norms for the following work are:

The first weeding of rice: twelve square '*brazos*'. One *brazo* is calculated by extending both arms and taking the distance from fingertips to fingertips. Each *brazo*, it is said, is approximately two metres long, and each hectare contains sixteen tasks.

The second cleaning of rice: if the rice is 'dirty' with weeds, the above measurement is used; if the rice is relatively 'clean', a larger strip of land is negotiated between worker and 'owner', a term described below.

Harvesting rice: the normal task is twenty *monotadas*, each of which contains three *puños*, a *puño* being a large handful of rice grains and stalks.

Cutting down the forest, cleaning scrub off the land, etc: the size of these tasks depends on the conditions encountered and is set initially by the owner and then negotiated with the worker.

In day work the labourer is paid for his time. The normal working day for hired labourers runs from 7.00 a.m. to 4.00 p.m. with an hour for lunch plus other rest intervals.

In return for his labour a worker receives cash, kind, labour, cooked food, or a combination of these items. Whatever the reward, the amount received for a task or a day of labour is the same: for example, for weeding rice a worker paid cash receives $1.00 for the task or for the day.

The owner proposes the unit of measurement, but he may have to change it in accord with the wishes of his labourers. The labourer usually prefers task work, for with this work form, the labourer sets his own pace, and by working steadily he often can finish in the early afternoon, allowing him time to rest, to escape from the sun, or to do some of his own work. If the task is a hard one, however, the worker may prefer day measurement, and of course the man who works fast to finish tires himself. The owner views the situation differently. If he hires by the task he is assured of having a set amount accomplished, but he has no control over the quality of the work, which may be important. Conversely, if the owner hires by the day the amount completed may be small, but the work may be of better quality. Much depends on the type of work to be done.

Weeding rice: usually this is done by the task, but if the weed growth is heavy or a variety of crops is seeded in the field, both worker and owner may prefer day work.

Seeding rice: this work is relatively easy and labourers are hired by the half-day. (The seeding is done in the morning as the rains come in the afternoon.)

Cutting down the forest, cleaning scrub off the land, etc.: normally this type of work is done by the day, since the difficulty of the activity varies by the plot.

Harvesting rice: this is done by the job or the day; if the owner is in a hurry, he may prefer task work.

The critical distinction between task and day work, thus, is that of quality versus quantity. In day labour the quality of the work is good but the owner cannot maximize the amount accomplished; in task work the quantity is set, but the quality may suffer. The alternative selected is determined by the wishes of workers and owner, the condition of the field and the task to be performed. But the unit of measurement also is set by the way in which the labour force is recruited, organized and rewarded.

The 'junta'

The largest and most spectacular form of organizing labour is the *junta* (*juntar*: to join, assemble, collect).[1] *Juntas* are formed primarily to accomplish jobs which require a large number of workers or which must be finished rapidly, such as cutting down the forest or scrub, erecting wooden fences, weeding crops, harvesting rice, uprooting and carrying a thatch house to a new location or building a new home.

The person who calls the *junta* is termed the 'owner'. He selects the men to come, and no one attends a *junta* unless invited. Usually the owner notifies the workers at least two weeks in advance. The *junta* lasts one day. In payment for the labour the owner must provide his workers with ample quantities of food and drink throughout the day. Ordinarily, *chicha fuerte* (the fermented maize drink) is prepared and sugar-cane alcohol also may be served. For large *juntas*, especially those held in the past, one or more hogs may be slaughtered. In addition, the owner owes back one day of *junta* labour to each of his workers. Such labour is owed only for the same task for which it was originally paid. If a man calls a *junta* to build a house, then he owes each of his workers a day's labour in a house-building *junta*: *junta* help in weeding their fields will not erase the debt.

To describe the *junta* in terms of abstract obligations, however, is to provide only a partial view of it, for in practice the work pattern is flexible and can be adapted to different conditions. In particular, the degree of obligation to return the work varies, and in fact *junta* types fall along a continuum of work debts. At one end of the scale is the large (from fifteen to seventy persons) house-building or agricultural *junta* at which food and drink are served and a definite work debt is incurred. A slightly lesser obligation to return the work is incurred by calling a small agricultural *junta*. At such a *junta*, often held the day before a *fiesta*, the owner provides drink and may serve food. Since it is difficult to recruit men for small agricultural *juntas*, anyone may attend without invitation. The labourers help the owner finish his immediate tasks but do not always work the entire day, as in the traditional *junta*. The men also take frequent breaks to drink, talk or *gritar* (a rhythmic shout exchanged between males). This form of the *junta*, then, sees a loosening of all the obligations. Sometimes a very small *junta* is called to complete a minor task, such as moving the roof structure of a thatch house. Such *juntas*, usually held on Sundays, a rest day, take but one or two hours. The owner explicitly requests the aid, but food and drink are not served, nor is a unit of labour owed in return. Lastly, for the 'co-operation *junta*' the recipient owes the workers nothing in return. In fact, he does not even organize the *junta* himself; rather, neighbours, kin and friends offer aid on their own. Such help may be given when a man is quite sick or has been thrown in jail, and his crops need immediate care. Although the beneficiary offers no food and drink, some of the workers may bring their own.[2]

The two extremes of the *junta*, then, contrast markedly in terms of their obligations. The owner of the normal *junta* is a provider, and in some cases has more goods than his workers. The recipient of the co-operation *junta* is in the opposite position: he is in need of comestibles. Indeed, so different are these work forms that it may be asked why the term *junta* is even applied to the latter situation. The co-operative group is classed as a *junta*, apparently because, like the 'true' *junta* and unlike all the other labour patterns, the immediate transaction is unequal, the work group is organized on the basis of individual not collective exchange and the labour may be returned, if at all, after a long interval. The social nature of the 'true' *junta* also is evoked by the co-operative work group. Of all the available words, the verbal category *junta* most closely fits the activity, but significantly this form of organization is lexically marked: the group is known as a *junta de cooperación*.

From the worker's standpoint the incentives to participate in a *junta* are slightly different from those he experiences in the other labour forms. In addition to its economic utility, a *junta* has an element of celebration. Nearly everyone recalls with pleasure some of the large *juntas* he has attended, and even a small half-day agricultural *junta* will be anticipated with pleasure. The festive character of the *junta* derives from the fact that drink is served, a convivial group is brought together, and the participants do not have to complete an assigned unit of labour. There are no specified hours of work, though most do come early and stay until the job is finished, and the owner may provide only general instructions for his labourers, who themselves make many of the technical decisions in the course of the day. In addition, the men customarily sing or *gritar* as they work; sometimes individuals or groups compete with one another to complete a set amount of work first; and a generation ago small dances were held at the conclusion of the activities.

The owner of the *junta* must possess some entrepreneurial skills. In the first place, agricultural *juntas* are called only by those who have a 'large' amount of land under cultivation. Since the land is held in usufruct, the size of an individual's plot is in part a reflection of his planning and working abilities. Holding the *junta* – securing the right quantity of food and making certain that it is prepared in advance - also requires organizational capacities. The *junta* permits and requires that a careful calibration be made between the work to be done and the force to be recruited, since the *junta* is a one-day affair. Lastly, in order to hold a *junta*, a man must have a spouse who can prepare the food; he is necessarily the head of a functioning, self-sufficient unit.

As a system of relationships, the *junta* is ego-centred. Brought together for one day only, the workers bear no obligations towards one another and may not even be acquaintances. Each separately enters into an exchange with the owner. The owner-worker bond, however, is long lasting, for theoretically the *junta* involves offering a day's labour for food and drink in exchange for the same, perhaps years later. Men ordinarily do remember their *junta* debts and credits, and I was able to elicit obligations which dated back twelve years. This delayed reciprocity itself has an economic advantage. If a man organizes a *junta* and recruits fifteen individuals to work on his house, he will be able to amortize his labour debts over time, for seldom will he be required to join several house-building *juntas* in the same year. From this perspective, the food and drink, besides their immediate value of stimulating the workers and enhancing the festivities, constitute a down payment for the labour and a promise that a final labour payment will be made on demand. Of course, over the years memories fade and debts are denied, but perhaps necessarily so, for otherwise it would be quite impossible for one individual to hold several *juntas* in the same year or even in proximate ones.

The 'peonada'

The term '*peonada*' is used by the peasants in two ways. Literally, the word means a crowd or group of peons. For the people, any group of eight to ten or more men working together in a field, which is not a *junta*, is a *peonada*. A smaller group, seven or less, could be termed a *peoncita* (diminutive form). But the word *peonada* also refers to a specific way of organizing a labour force. It is a semi-permanent group which works the fields of each of its members in rotation.

The *peonada* organization is deployed only for the larger and heavier agricultural tasks, such as cutting down the forest, or weeding and harvesting rice. A *peonada* is organized about a week in advance of a job, but the same group may continue to co-operate through the year on other tasks as well. Each member contributes equally to the group: the exchange is one of labour for labour. Generally the work given and received is for the same job, and the entire group decides in advance whether members will work by the day or by the task. If an individual cannot attend on a particular day, he should send a substitute. At the end of the agricultural year any outstanding debts among the members are cancelled through cash or individual labour payments.

Most *peonadas* do not contain more than twenty members, a dozen to fifteen being the usual number. Such a group seems optimum; it is able to complete in a day the larger tasks of a one-hectare plot, while a bigger group may have difficulty exchanging labour equally within the year. The *peonada*, however, does not have an inflexible organization. If a man's field is overrun by weeds, the group may work there for several days in succession; or sometimes it splits into smaller groups of three or four so that several fields can be started at once. The order of working the fields is determined at the outset, and those who join last are placed at the bottom of the list. But everyone has some concern that his own field may not be worked in time.

For the participants the *peonada* offers certain advantages. If a man contributes labour, he can be certain of having some of his arduous tasks performed for him, without having to pay cash or goods. Working in a group and having the support of others when facing heavy tasks is also a reward for many.

Compared to the *junta*, the *peonada* is a 'utilitarian' form of organization; its incentives are more specific. Food and drink are not normally served by the recipient. Each member is motivated by the knowledge that his own crops will be adversely affected if he performs poorly for others. Yet, the *peonada* in practice sometimes differs from theory. For example, a *peonada* recipient occasionally serves his workers *chicha* to express his goodwill and increase their motivation. This deviation, however, raises a problem of classification, for in casual conversation, at least, a person will say that if *chicha* is being served, then a *junta* is being held.

The *peonada*, unlike the *junta*, is deployed only in agriculture, and it is not even utilized for all farming tasks. Furthermore, the *peonada* does not permit an equilibration to be made between size of work force and magnitude of task. In the *peonada* the group size is determined first, and a man must then find enough work in his own fields to occupy the group, if he wishes to get a full return, or he must recruit further labourers himself, if he has yet more work to be done.

The internal organization of the *peonada* also differs from that of the *junta*. The group is bounded and works together for a set number of days. Reciprocity is neither immediate nor long-term; remuneration is not paid at the end of a day's work, yet the labour debts do not stretch beyond the agricultural year or set of tasks.

Earned peons

A direct exchange of labour between two persons is termed 'earning a peon'. The work may be by the task or job, and the exchange usually is completed within a week or two. But each participant may form a similar contract with others, and thus a man may work for several days gaining peons in different places in order to collect all the debts on the same day. In such a case the co-workers have no mutual obligations. Unlike the *peonada*, earning a peon does not involve an organized group, and the debts are paid more quickly.

Flexibility is one advantage of this labour form. An individual can recruit a labour force to match his needs and he does not have to possess cash or goods to obtain the help. Fellow labourers can be selected carefully, poor workers may be avoided. Also, the organizational form can be used for nearly any task: two women may exchange domestic help after each has given birth.

Although earned peons works well when labour needs are minimal, size limitation is its principal disadvantage. If a man needs the help of four or five others, he finds it complicated to organize working for each and then collecting the debts all at once.

The host of the peons is termed the 'owner', and an owner of a labour force, regardless of organizational type, is obligated to provide his workers with a sharpening stone for their machetes. The owner also must supply his labourers with water. Usually he brings a large gourd which he fills at a nearby stream, and either he or perhaps a child takes the jug around to the men. As the people explain it, the custom of the owner providing water and a sharpening stone is a matter of efficiency, but the practice has also symbolic overtones. An owner who has not provided his workers with drinking water and a stone will be sent home to secure them - as if these were the minimal requisites of office, to be elaborated at times into food and drink.

Hired peons

The form of labour most frequently used now and which applies to the broadest range of activities is hiring a peon. This organizational type may be used for all agricultural work plus such activities as mending fences, tending cattle or building houses.

Hired labourers are paid in cash or kind. For most work the wage rate is $1.00 to $1.25 for the day or the job. The payment must be made directly after the work is completed or even before - the classical wage advance - so that the worker can buy enough food to sustain himself. In times of cash shortage a man may have to sell a chicken or other good to pay his workers promptly.

Only infrequently is the hired labourer paid in rice or maize, such payment usually being made during the harvesting season. Remuneration in kind is worth slightly more than in cash. In cash the rice harvester is paid $1.00 for a day's work or $0.05 per *manotada* cut, assuming that he will harvest about twenty *manotadas* per day. In kind the equivalent payment is four *manotadas* per day, and while the size of the *manotada* varies, each *manotada* is worth from $0.25 to $0.30. Financially, there is an incentive for the worker to ask for food payment. Owners, however, are reluctant to recompense in kind unless they have an abundant harvest or are short of cash, while receiving payment in kind is a sign that the labourer himself is in need of food, an embarrassment since it reveals that he is not a self-sufficient producer. Thus, an owner not infrequently must rush a portion of his freshly harvested subsistence crop to a local mill to obtain cash for paying his fellow subsistence producers.

One obvious advantage of this form of mobilizing labour is that it requires immediate payment, usually cash. Instant cash remuneration frees the owner from further obligations toward his workers, while providing them with choice in their ultimate reward.

The owner usually works in the fields with his hired labourers, and he is able to exercise greater control over them than he can in the *junta* or *peonada*. Hired workers are expected to arrive at a specified hour; the tardy one may be refused work. The owner describes the day's work to be done or divides it into tasks. As the day progresses the owner may offer a few suggestions about the work, although he must be careful, for the slighted labourer may quit. Usually, an owner will bear with a poor worker, but then not hire him again; indeed, most people know about the work habits of their fellow villagers. Unlike the *junta* the individual's work capacity rather than his total person is being engaged; hired peons represents alienation of labour from human.

When the owner himself cannot be in the fields, he may hire a *mandador*, an intermediary to oversee the work. After receiving general instructions from the owner, the *mandador* hires the workers and performs all the other tasks of the owner, such as supplying water and dividing up the work. The *mandador*, the foreman, acting on behalf of another, seemingly can order the peons even more directly than the owner himself. He is paid slightly more than the rest of the workers and does not undertake actual physical labour. At the end of the day, however, the owner comes out to the fields, views the work, and pays the workers.

The size of a hired labour force varies, although seldom is a man able to recruit more than a dozen workers for a single day. Groups of six or seven seem to be the maximum, and if an owner has much work to complete he may employ the men for several days in succession.

Although under greater control of the owner, the workers do have certain freedoms. They may demand to be paid in cash or kind, and they can bargain with the owner about the amount of work to be done and whether it is to be measured by the day or by the job. With task work the labourers themselves decide who undertakes which job, and sometimes a man inquires in advance about who else will be working to know if the load will be shared equitably.

Contract labour

The final work pattern, which combines elements of the preceding, is contract labour (*el ajuste*). This form is used for all agricultural purposes except the harvesting of sugar cane. A man offers a contract when he is unable to work in the fields himself and when he has at least five days of work to be completed. *Ajustes* usually fall in the range of $10 to $20, or ten to twenty days of labour.

To form a contract the owner contacts another man, who then

decides whether to assume sole responsibility for the contract or to share it with others. In the former case, the contractor views the work and bargains with the owner; he, then, hires others to do the labour, paying a normal wage rate. The contractor may or may not work in the field with his labourers. Alternatively, the contacted person may invite several others to join in the contract, in which case each views the field and must first agree to the overall payment offered. The contract bearers do the work and split the total payment. Since it is difficult to calculate the precise time a job will require, the people say that gains and losses are made in an *ajuste*. Using $1.00 per day as the base rate, they figure that if an owner offers $14 and the job is completed with ten days of labour, then a profit has been earned; the reverse creates a loss for the contractor(s). Thus, when a contacted individual sees that the contract work can be completed quickly, he usually takes sole responsibility for the job. Conversely, if the initial man thinks a profit can be realized only if all the labourers work hard, he may offer to split the contract with others.

The owner bears no responsibility for the job once the contract has been made. He does not work with the labourers nor does he have to provide them with water and a sharpening stone. He is assured of having all the work done at a specified price, and his cost may be less than if he hired workers on an individual basis. On the other hand, the owner has no control over the quality of the work; frequently, the labourers work rapidly to make a profit in addition to their wages.

The *ajuste* is not viewed by the people as a separate form of organization, comparable to the *junta*, *peonada*, exchanging peons, or hiring labour. Conceptually, it is considered a variant of hiring labourers, and it is sometimes said that task work lasting for more than a day is an *ajuste*. The position of the contractor also resembles that of the *mandador* for hired workers.

The forms compared

A comparison of the four modes of organization reveals more clearly the principles upon which each is based. Overall, the patterns are imbued both with the dynamics of a subsistence system which is contained within capitalism and with the contrary forces which exist within subsistence itself.

Hiring peons, unlike the three other forms, represents the impress of the market system upon the rural, subsistence economy. As a transaction, hired peons consists of an inequivalent exchange of cash or raw food for labour. The *junta*, by contrast, is theoretically a mutual exchange of processed food, drink and labour for the same, while both the *peonada* and earned peons consist of an exchange of labour for labour. The latter three forms limit the recipient to a reward which can be used in the subsistence sector alone. The duration of the transactions also differs, for hired peons are remunerated immediately, but earned peons are repaid after several days, the *peonada* is completed only within months, and the *junta* exchange may take years before being brought into balance. As delayed transactions the final three forms require an element of trust and are most fittingly formed between acquaintances or fellow villagers, that is among subsistence producers.

Owner and worker also are more clearly separated in the hired peon transaction, and the paid labourer is held the most accountable for his task. *Mandadores* - in capitalistic terms, foremen - occasionally are used for hired peons, and the *ajuste*, the contract variant of hiring peons, is explicitly calculated on the basis of profit and loss.

More important, there is a difference in participation between hired peons and the other forms, for villagers who work as hired labourers are most frequently younger males - men who are unmarried and still living in their fathers' houses, or men who have just separated from their natal homes. The self-sufficient household head does not usually become a hired peon. As the people point out, it is not the 'custom' for the older ones to earn money: 'The old ones don't leave if there is a son to go.' If, however, an older man does not have a current crop, has insufficient crops or has a poor crop, he also may enter the local labour force. To speak of class formation or incipient stratification in the village would misrender the reality of everyday life, but the distinction in participation between hired peons and the other forms *is* based on differential access to property. Those who control resources hire those who do not, a fact which fits the earlier observation that the greatest potential for exploitation lies within the father-son relationship. Hiring labourers offers the possibility of realizing surplus labour.

On the other hand, while the transaction of hiring peons does emanate from the market, the pattern itself is traditional. Even the oldest people recall that as youngsters they earned cash as field labourers. The form of organization does not represent a recent change but the 'containment' and evolution of subsistence within capitalism. Significantly, hired peons may be paid in cash or in kind, non-market goods. Even the use of money as remuneration does not in itself imply that the transaction derives entirely from the market. The wage paid actually is below the value of the labour performed (see Table 5); yet this fact is not seized upon by the people to make subsistence agriculture into a profit-making venture. Rather, the remuneration serves negatively as an incentive to devote time to one's own crops. Neither fully one nor the other, hired peons is a transaction poised between subsistence and capitalism, though it leans towards the latter.

Although the *junta*, *peonada*, and exchange of peons are all subsistence sector transactions, the *junta* differs significantly from the other two. Only in the *junta* must cooked food and drink be served. Raw food is produced in the fields and paid to hired labourers, but cooked food, which comes from the domestic domain, is served to favoured guests, and a person so honoured ought not to refuse, for to do so is a sign of enmity and distrust - the refuser indicates he believes the food is poisoned. In the *junta*, then, cooked food is extruded from the household back to the productive domain, conveying with it the conviviality, commensality and closeness of the home. Host and workers eat together in a friendly and trustful way.

The festivities of the *junta*, however, can serve to disguise what are potentially disruptive relationships among individuals. Economically the agricultural *junta* serves well where productive differences among persons appear. For the owner, the *junta* is an effective means of converting 'excess' consumption items into productive labour. To a small degree this yields 'prestige', a non-indigenous term, for the people are able to name those who in the past held large or good *juntas*.

From the participant's standpoint, the worker is expressing his willingness to consume part of his reward immediately, his lack of need to carry raw food home, and his ability to postpone the eventual reward of receiving labour help. The *junta* participant shows that he, too, is a self-sufficient farmer who even has time to give to others. If a man has no crops or is unlikely to build a house, he has little incentive to join a *junta*, which means that the *junta*, unlike hired peons, holds little attraction for young men. On the other hand, in practical terms, those who join a *junta* often do so in order to receive an immediate consumption item - food - knowing that they may never need or be able to collect the productive, labour, part of the return.

The *junta* thus realizes one tendency of a subsistence system. In theory, it is an equivalent exchange among equals, but in fact it may be an unequal exchange of useful objects for labour, of goods for labour power. The *junta* provides a means for converting dead labour to productive labour. Yet, it is neither a capitalistic relation nor sustainable. Unlike hired peons the owner provides not only 'necessary maintenance' to his participants but a surfeit, a celebration, otherwise no one will attend, and thus the owner derives little surplus. The labourers also are held less accountable for their work than are hired peons. Although theoretically it might be possible to convert a product surfeit to productive labour to product to labour over time, the *junta* is neither a permanent relation of production nor in fact accumulative, given the size of payment and vagaries of subsistence production. Thus, the *junta* is contained within and does not in itself lead to the destruction of subsistence production. On the contrary, under the cover of conviviality, of processed comestibles which come from the home, the *junta* converts possibly destructive wealth disparities into bonds of closeness and sociability. What the act of hiring peons says openly, the *junta* hides behind the screen of resource squandering. It is a means for managing the contradictions of productive differences and short-term accumulation which appear in a subsistence economy.

Like the *junta*, the *peonada* is a large group of workers who themselves have agricultural jobs to be done; young men without crops have no incentive to join.[3] But unlike the *junta*, the *peonada* is most easily formed among those who have similar sized fields and similar crops, minor ecological variations permitting the work to be staggered without difficulty. Participating in a *peonada* does not give one prestige nor allow for the conversion of goods to labour. Only the *peonada* consists of a closed group of men exchanging their labour equally. The members tend to be friends and in this respect the *peonada* represents solidarity among subsistence producers.

The pattern of earning peons differs from the *peonada* in that, like the *junta*, it is an egocentric exchange, consisting of a sum of contracts between an owner and others. It is the most 'utilitarian' of the subsistence forms in that it permits a man to recruit labour when he needs it or to 'bank' his surplus time for the future. In practice there is no clear distinction between the small *peonada* - the *peoncita* - and earning peons. I knew of instances, for example, in which three or four men reciprocally exchanged labour among themselves. They labelled their activity 'exchanging peons', but admitted that it was similar to a *peonada*. Their use of the expression, 'earning a peon', they justified by pointing out that if a participant were unable to appear on a particular day, instead of directly sending a substitute he might pay the debt at another time.

The *peonada* and earning a peon are true equivalent exchanges of labour for labour. Representing the equivalence and inter-dependence of production units, they do not compromise the fundamental internal organization of the household nor emanate from a form of production other than subsistence. Based on the assumption of household self-sufficiency, they answer certain requirements of the agricultural system and realize a limited form of cohesion through the exchange of that most valuable asset, productive labour.

Expressed formally, the *penada* is a symmetrical exchange of labour for labour within a group, and earned peons is a symmetrical exchange of labour for labour amongst individuals. The *junta*, in theory, is a symmetrical exchange of labour and processed food for labour and processed food among individuals, while paid peons is an asymmetrical exchange of cash or kind for labour among individuals. In practice, an individual may combine the different modes in one operation, using for example both hired and earned labourers, paying in cash and kind, or augmenting a *peonada* with paid peons. But current practice also differs from the traditional due to the transition towards a market based economy.

Desuetude and disarray of the forms

In recent years the frequency of use of the different forms appears to have changed dramatically. The last major *peonadas* were held in the year immediately prior to my arrival, and though in 1966 I observed one *peonada* of six people, there was little doubt that the *peonada* would never again be practised. Large harvesting *juntas* have not been held for several years, although smaller house-building and other *juntas* still are. The exchange of peons is carried on, but the predominant way to recruit labour now is by means of cash payment.

The example of one *junta* held to build a mud home illustrates some of the reasons why the older forms of labour exchange are both little used and are falling into disarray under the impact of the market orientation. In Table 8 are listed the costs, both of labour and in cash, which the owner incurred to build his home. This was, in fact, a comparatively small *junta*. The owner pointed out that his father had built a comparable house in 1940 but had gathered over 100 people to assist him; he, however, by doing many of the tasks before and after the *junta* - often receiving a few moments of aid from friends - had cut down the amount of work to be done on the day of the *junta*. Also, many of his supplies were purchased, whereas previously they or their equivalents could have been supplied from a food surfeit in the household or secured for free in the countryside. Most noteworthy, however, was the reward which he gave his labourers. He collected nine old *junta* credits in the traditional fashion but was unable to find anyone who would offer a pure *junta* debt; to recruit more help he was forced to offer a wage as well as the perquisites of the *junta*. The result was that this work form, which was typical also of other house-building *juntas*, did not fit the traditional categories. Thus, when discussing and classifying the day's activity the owner termed it a *junta*, but others claimed that he really had hired labourers and had added food and drink to stimulate their work. Finally, one man, when taxed with the problem of defining this new labour form, invented a neologism: for him the labour group was a *peones ajuntados* or a *junta apeonada*, his linguistic expression mirroring the anomalous nature of the activity. But it appears doubtful that the anomalous form will become standard practice. Currently in a state of disarray, the *junta* will not long endure.

The causes of the changes in the labour patterns may be sorted roughly into those emanating from within subsistence and those deriving from the market, although the two facets are not independent. The decline of the thick *montes* and subsistence harvests has had several impacts. There is less need to recruit help at certain stages of the agricultural cycle, and people have fewer surfeit goods to provide *junta* labourers. As evident in the example, the costs of holding *juntas* also have risen in that now the comestibles served

TABLE 8 A house-building *Junta*

Tasks and materials	Owner's labour	Cash outlay
I Prior to the *junta*		
Securing the support poles	2 Days	
Securing thin cane strips	3 Days	
Securing vines for binding	2 Days	
Cutting vines to size	2 Days	
Securing thin wood for the lattice	2 Days	
Transporting the thin wood by oxcart	1 Day	$4.00
Emplacing the support poles	1 Day	
Constructing roof frame	1 Day	
Adding lattice to roof	2 Days	
Wood for door frame		$3.50
Making door and window frames	1 Day	
Putting up lattice for walls	2 Days	
Loading and transporting dirt for walls	1 Day	$4.00
Cutting roof thatch		$5.00
Transporting thatch in cart	1 Day	$1.00
Cutting and transporting straw for mixing with dirt	1 Day	$2.00
Two pounds of nails		$0.30
First-Stage Totals	22 Days	$19.80
II The day of the *Junta*		
Cart to bring water for mixing with dirt		$4.00
Three paid peons		$3.75
Nine peons previously earned in *juntas*	9 Days	
Meat		$4.00
Molasses for *chicha*		$1.50
Six pounds maize for *chicha*		$0.30
Making *chicha* (the day before)	1 Day	
Fifteen pounds rice		$1.88
Sugar-cane alcohol		$6.00
Second-Stage Totals	10 Days	$21.43
III After the *Junta*		
Five sheets of corrugated roofing		$8.50
Used tile for roof		$10.00
Putting on tile and corrugated roofing	.5 Day	
Putting on thatch	1 Day	$1.25
One sack cement		$1.70
Finishing and cleaning up	1 Day	
Third-Stage Totals	2.5 Days	$21.45
Total Costs in Labour and Cash	34.5 Days	$62.68
Total Costs Converting Owner's Labour at $1.25/Day to Cash Equivalent	$105.80	

must sometimes be purchased. In addition, the pressures on subsistence production have made it more difficult to arrange labour exchange on a large scale. Now, more than before, plot sizes in subsistence differ so that it has become difficult for individuals to find others with equivalent labour needs. And the people have less unused time now, as a result of the decline of the surplus. Also, with harvests diminishing no one wants to be the last aided in a labour group nor to host a festive group that may be careless or not finish the appointed job. As before, this congeries of forces leading to changes in labour organization may be termed factor push.

But the elements comprising market pull - the desire to be 'civilized' and leave aside old customs - have had an equal if not larger impact on the labour forms. The cash crop has had a profound influence on the labour situation, in addition to its effect on the subsistence crops themselves. The growing of sugar cane has led to an absorption of community labour. Individuals who grow sugar cane must devote part of what was their surplus time to the crop, time which facilitated the formation of the subsistence labour forms. In fact, during the harvesting and weeding periods of the sugar cane, labour must be imported to the community. Labourers from distant areas, however, do not want a work credit but cash in order to buy market goods. Just as important, sugar cane is raised for cash profit and given the different sized fields which people have, neither the equivalency of the *peonada* nor the squandering of the *junta* is an appropriate way to mobilize a community work force in the cane. Furthermore, with the diminution of the traditional crops and the breakdown of full subsistence provisioning, stemming from the intrusion of the outside economy, even spending a day earning a peon in subsistence is less attractive, not only because of one's own lack of surplus time and lesser need for subsistence agricultural help, but also because of the necessity and desire to have money for purchasing market items. Like outsiders, village members increasingly want to receive cash wages.

To suggest that the desuetude and disarray of the traditional labour patterns is due solely to the superior 'flexibility' which cash affords over a use value or to the lifting of 'traditional ascriptions' would be not only to assume an evolutionary viewpoint but to confound description with cause. The clash of subsistence production and capitalism has had its inevitable impact on the forms of labour organization.

Yet, this change is not without its lessons. Within the subsistence system, labour exchange – in whatever form – provided one counterbalance to the 'centrifugal' (Sahlins, 1972: 97) tendency of a household economy. Mistaking, perhaps, result for cause, the peasants observe that they have fewer *peonadas* and *juntas* now because they have less 'union'. As the rural area moves toward greater reliance on market principles there is a transition from household to household integration via labour exchange to household-market integration through the hiring and sale of the person's labour capacity. The fundamental units of the society are not changing but their focus of integration is, from the community to the market. This fact suggests once again, although from a different perspective, that the traditional forms of inter-household labour organization are not themselves fundamental relations of production. Answering certain agricultural requirements, they are and were primarily manifestations of forces and contradictions generated by the subsistence system itself. With the undermining of this system the contradictions and forces are being muted.

6 The seeds of a transformation

Sugar cane has long been seeded in the countryside for immediate use, but the raising of sugar cane for exchange is recent. With the completion of the concrete highway through Los Boquerones in June 1958, raw cane could be transported economically to the two mills which lie toward Panama City. As these well-established mills were desirous of expanding production to meet Panamanian and world demand, in the late 1950s they began to extend their sources of supply to include Los Boquerones and certain neighbouring areas. By the early 1960s the first harvests from the village were being milled, and each succeeding year has seen more *campesinos* planting the crop and in ever greater volume.

The primary products of the mills are a consumable item and partially refined brown sugar which is exported to the USA. The side products of molasses and alcohol also are sold, while the squeezed and dried stalks, known as *bagazo*, provide fuel for the factory. Sometimes the peasants buy, with the money they have earned, the product they have helped to create.

This rapid acceptance of sugar cane as a cash crop within the community is both a simple and complex transition. On the one hand, with few changes in productive techniques, the peasants have been able to integrate the growing of sugar cane into their customary work cycle; sugar cane supplements the traditional crops. But with the planting of sugar cane the peasants now are producing a raw as opposed to a finished product, an item which only can be exchanged for something else as opposed to a good which can be consumed. That the new crop is sugar cane and not some other vendible harvest is of little importance, for only an external, roundabout tie links producer and object. Cane has 'utility' only when the peasant rids himself of it, the reverse of that which he does with rice. More profoundly, the sugar cane triggers a change in the relations between peasant and national economy. Through the sugar cane the people reap the advantage of becoming national consumers; as they are 'integrated' into the national economy, however, they lose control over their productive means. Thus, this minimal shift in productive techniques actually represents and leads to a total economic transformation, for the peasants change their worldly conditions, from being independent, self-sufficient producers to becoming petty capitalists and day labourers.

Producing sugar cane

Peasant sugar-cane producers are termed '*colonos*'. In theory, though not always in fact, sugar cane is planted in a field which is in its second year of subsistence use. The rice and maize are first seeded and weeded, then the cane is added, usually in June. Planted in this fashion, it will have enough time to grow but not so much time that it will come to dominate the subsistence crops before they can be harvested in August and September. Seeding the sugar cane, however, does close off the possibility of planting a long-maturing rice or a second maize crop.

For each hectare to be planted about 1,500 cane stalks are required. The stalks are cut into pieces, each piece containing several eyelets or seeds. Using the steel-pointed digging stick, the producer makes angled holes in the ground about a metre apart, and each cane piece is slid into a hole until only a small portion remains above the earth. Eventually each such seeding will produce a clump of cane, which itself will throw off sprouts each year. In July of the first year as the rice and maize receive their second weeding, the sugar cane receives its first. The physical process, using machete and curved stick, is the same for all three crops.

After the rice and maize are harvested their stalks must be cleared from the field. Effectively, then, the cane receives a second weeding in September, October or November of its first year. But this cane, which is known as 'eighteen-month cane', is not cut in the following harvesting season which runs from December until April. It must be left to grow for a year and a half. Thus, a year after seeding, usually in June or July, the cane field is given yet another thorough weeding.

In the second harvesting season, or *zafra*, the field is ready to be cut. Each stalk is sliced at its base and thrown on to a small pile. The cane truck is driven on to the field, and the workers shoulder the cane, climb a ladder and load the stalks on to the truck. Work in the cane is never easy, for the cane is hot, but the men say that loading a truck is one of the most arduous and tiring of the tasks.

After the first harvest the cane is known as *caña de cepa* (stub, stump). *Cepa* cane is weeded in the June or July following its initial harvest; but after this no further weedings are given. Stump cane yields a profitable harvest for two to three years, but then its sugar content drops, and the land itself in the absence of fertilization becomes exhausted. A cane field too old to harvest is a rather barren looking spot.

Harvesting can be accomplished more quickly and with less labour, if the field is first burned, for then the workers need not strip each stalk of its leaves. But only a field which previously has been harvested is strong enough to withstand the burning. Once burned the cane must be cut and sent to the mills directly, for it rapidly loses some of its sugar content. Because of the potential for loss, the mills do not receive burned cane unless they have given prior permission and can grind it immediately.

When a field is not burned, the stripped leaves left on the ground inhibit weed growth but also the new cane, and thus a field unburned before harvesting sometimes is burned afterwards, particularly when the dried leaves present a fire hazard.

About fifteen different strains of sugar-cane seed are known; some are traditional varieties good only for home use, while others have been introduced by the mills, which maintain experimental stations. The types vary in weight, thickness, sugar content, durability, extent to which they throw off sprouts, suitability for different soils, disease resistance, and ease of harvesting. But compared to the rice the people have less knowledge about and control over the cane seed they plant. There is uncertainty about which type of seed should be used in different areas, and some of the newer strains are not available in sufficient quantities for massive planting.

In addition to the decisions about burning and seed type, a third has to be made concerning timing of the harvest. In general, sugar content in the cane peaks in March and April, toward the end of the season, and by this time the cane also has had more time to grow. On the other hand, if a man waits too long in the *zafra*, the rains may come, making the feeder roads impassable by truck; indeed, the harvesting season opens and closes with the rains. Also, fewer workers are available late in the season, as many of the outside labourers return to their villages to work on their own crops. As an inducement to harvest early, one of the mills has a special two-week period at the onset of the harvesting season during which it pays the top price for raw cane, regardless of sugar content. Within these parameters a man may want to harvest if he is in need of cash; or, if his field accidentally has been burned, the mill will permit him to send the crop immediately. But usually the specific timing of the harvest is set by the trucker, who has to spread the cane harvest of his suppliers evenly over the season. This, of course, limits the individual *colono*'s chances of optimization. In fact, some producers lose considerable time searching for or dispersing labourers when they are told suddenly that they must harvest or cease harvesting; and not infrequently some *colonos* find themselves at the end of *zafra* with uncut cane. These latter are only two of the risks that the peasant must assume in order that the mills may be assured of a steady supply of raw cane.

As with the rice and maize, sugar cane can be raised in variant ways. The most important of these permutations involve the use of new technology. Both mills are trying to introduce fertilizers, herbicides and even ploughing, but these techniques are being accepted on a haphazard basis. A primary problem is that the people have inadequate knowledge about the new methods. One man, to cite a single example, borrowed a hand sprayer and secured a herbicide from one of the mills to use on his recently seeded cane. He was told not to spray the young cane immediately, for this would kill the crop, but, as he explained to me, he did not quite believe this information and went ahead. The result was total loss of his seeding.

Labour in the cane

The single person cannot raise sugar cane alone. For the seeding, weeding and harvesting a group must be gathered, even for a small field. Sufficient labourers for the seeding can be recruited from within the community, but for weeding workers often are found from without, while harvesting always requires that recourse be made to the labour market.

In Santiago there is a square block marketplace at which goods most in demand among the *campesinos* are sold. One corner of this covered area serves as the marketplace for labour. The labour market is open one day a week; and here, late on Saturdays during the harvesting season, gather peasants and Indians from non-cane, usually distant, areas who are looking for work. Such persons are easily spotted by the machetes and sacks of clothes which they carry. Both the *colonos* and the mills do their principal hiring in this market.

Relative to the mills, *colono* producers find themselves at some disadvantage in the labour market, for the corporate organizations can offer more elaborate work inducements. They provide transportation to the mills in company buses, and shelters where the workers may live. While the mills pay only $1.00 for each ton of cane cut, all their cane is first burned, making the cutting quicker. Mill sugar cane also need only be stacked on large chains, which are tied and then mechanically winched on to trailers for transport to the factory. Each worker is able, therefore, to earn between $2.00 and $3.00 a day at the mills, which is more than is offered by the *colonos*.

Although many of the outside labourers prefer to work for other *campesinos*, the villagers are finding themselves in an increasingly competitive situation. Many report that they have to detail in advance where the work will be, what kind of shelter is available, where firewood and water can be obtained, and what condition the cane is in. In effect, wage workers are demanding of the *colonos* more fringe benefits. Occasionally, a villager returns from Santiago alone, even when he has agreed to send his cane to the mill in the coming week.

This pattern of hiring also is inflexible and inefficient from the *colono* standpoint. Labourers from Santiago must be hired for five or six days straight, since the market is not open during the week. The mills are large enough so that they can always shift a labourer from one task to another, but the small-scale producers do not have this capacity. Thus, *colonos* sometimes are caught in the situation of having to find other, non-cane, work of their own, or other *colonos* desirous of labourers, if their harvesting does not require a week. Alternatively, a *colono* may have to hire a small labour force and work it quite intensively.

On the other hand, use of the labour market does have its advantages. Lump sums are advanced from the mills to the *colonos* to finance the weeding and seeding, but wages are paid in arrears for the harvesting. A village labourer, when he becomes disgruntled with his employer, will not show up for work, but Santiago labourers of necessity work for the week. The *colonos*, like the mills, have greater control over an outside, paid work force, for in such a situation the individual's work capacity is definitively separated from his social *persona*.

The sugar-cane mills

Both mills are 'factories in the field' (McWilliams, 1939). Located in the neighbouring province of Coclé, one is about thirty minutes from Los Boquerones, the other approximately forty-five minutes. One of the mills grows about 35 per cent of the cane it processes, purchasing the rest, while the other grows 70 per cent of its supply. Aside from this difference the two mills are similar in organization and aims.

In 1965 one of the mills had total dollar sales of approximately $3,500,000. It produced 41,000,000 pounds of sugar, of which a sizeable portion was exported to the USA. The mill purchased 153,963 tons of raw cane from *colonos* and raised 75,663 tons on its own farm to make a total of nearly 230,000 tons which it processed. But much of this growth is recent. In 1961 the company milled only 130,000 tons. Its output has nearly doubled in five years.[1]

According to the mill, its average independent producer supplies about 200 tons of raw cane each season, although it also claims to have 1,000 peasant producers. In 1965 the mill paid $980,000 to cane producers and truckers; of this amount the transporters received $211,000, or 21.5 per cent of the amount paid for the raw cane. The average producer, thus, received $6.36 per ton before transport cost. If, as the mill states, the *campesino* producer usually harvests 40 tons of raw cane per hectare, then on average each of its *colonos* has under cultivation 3.85 hectares.

The production figures are somewhat different for Los Boquerones where transport costs are higher, ranging between $2.00 and $2.50 per ton, and where 40 tons per hectare is not the norm when hand methods are used. The mill, employing capital intensive techniques, is able to achieve on its best land a yield of about 100 tons per hectare; some of its established *campesino* producers utilizing more productive techniques than the people of Los Boquerones also exceed the 40 tons per hectare figure.

The number of people employed by the mill varies. Throughout the harvesting season, 300 people work in the factory, but this number shrinks to 180 from June to December. During *zafra* 15 employees operate the testing laboratories, a number reduced to 2 or 3 in the off-season. On the mill's farms during the harvest at least 500 peons are needed to cut cane while 300 more operate field equipment. About 30 people are employed throughout the year in the local administrative offices, while 10 more run the firm from its Panama City office. The owning family occupy all the top administrative positions.

Day-to-day running of the mill is under the management of a professional administrator, who also provides liaison between the Panama City office and the factory. Working under him are the various functional division heads of the factory plus the field superintendent. The field superintendent oversees cane production on both mill and *colono* land, and under him are both a field chief for mill land and a *colono* chief for *campesino* producers.

The peasant suppliers have been organized into four geographic zones, each containing 250 *colonos* and under the direction of an inspector. An inspector earns about $100 per month and has use of a car. Problems the inspector cannot resolve are referred first to the *colono* chief and then the field superintendent. *Campesinos*, thus, have formal contact with the mill through an inspector and sometimes through the *colono* chief and field superintendent. They have little or no knowledge about other aspects of mill operations.

The local staff is aware that *campesinos* are not profit-orientated farmers, but aside from voiced concerns little is being done to aid the peasant. As one stated,

> We realize the *campesino* cannot only cultivate cane, he must also grow rice, maize and beans. But the problems are big, and we are limited in our resources. Within our limitations we are introducing new agricultural practices and giving economic help to the *campesinos*.

The mill with its efficient machinery is not yet operating at capacity. If, for example, the harvesting season is calculated at 120 days, then the mill, which can grind 2,400 tons daily, was in 1965 operating at under 80 per cent capacity. Given this processing potential, the world market and Panama's sugar quotas, the mill's first and stated objective is to increase raw cane production. Since it already is working its own land intensively and does not want to buy more, the corporation is focusing its efforts upon increasing *campesino* production. With its recent expansion into new communities, however, the mill does not want to augment either its number of producers or supply area, beyond the current twenty-mile radius. The cómpany believes enough land is 'held' by its present *colonos* to yield sufficient cane, if agricultural productivity is improved. The mill, therefore, wants to introduce the use of more mechanization, fertilization, and irrigation among its peasant suppliers.

The extension of technology and credit are under the guidance of the four inspectors but technological change has been slow, and most of an inspector's time is spent invigilating the loans. Cash is extended to finance seeding and weeding; fertilizer and other materials may be secured on credit. No interest is charged on the loans, but all are deducted by the mill before the *colono* is paid at harvest time. The amount loaned varies but is usually in line with the work proposed. If, however, a man has entered a poor harvest he may not be extended credit in the following year.

From the *campesinos*' standpoint the loans are low risk, as the mill attaches only the cane harvest, but high risk in that crop failure may leave the producer in debt for some time to the mill, which can make certain that the loan is repaid before the producer's profit is taken. Indeed, the *colono* bears a certain amount of risk for the mill itself, since a factory malfunction, which halts grinding, can leave the producer with peons and harvested cane, which quickly loses its sugar content.

The mill is concerned that money extended for weeding in June and July often is not directly invested in the cane but diverted to household uses. Such cash does arrive at a propitious time, for it is in these months that the subsistence harvests are ending, even earlier now with the advent of the sugar cane itself. In so far as the *colono* himself does the field work, use of the loan to meet household needs is, of course, appropriate.

The mill also has broader social impacts on the life of the peasants. It constructs and maintains feeder roads to the fields, charging a portion of this work to the *colonos* affected. Problems of individual producers may draw the mill's attention, such as boundary disputes. When a fire in one man's field damages the cane in another's, the mill itself may assess damages and transfer an appropriate amount from the account of the one at fault to the other.

Despite the many problems which the mill perceives that it has with its suppliers, it has been able to effect a rapid change in the agricultural system of the *campesinos*. The mill's short-term goals follow from the particular conditions in which it finds itself, but its overall strategy is that of a growing capitalistic organization: expand into new markets, augment production and increase productivity. In a most direct way capitalism is having a profound impact on the economy and people of Los Boquerones. The physical sign of this intruding system is sugar cane.

Transportation of the sugar cane

The mill claims that the transport system for the harvested cane is independent of it, but while this may be true in theory, it is not in fact. Informally, the inspectors often check on a *colono* through his trucker. Formally, the mill regulates the supply system through the transporters.

One hundred trucks service the mill whose principal supply of cane comes from *colonos*. Ninety-two of the trucks are owned by *colonos*.

The price of a used ten-ton truck begins at $3,500. The mill usually loans the trucker a part of the initial capital, the rest being secured through a loan from the sales agency. With his revenue, the trucker pays off the mill and makes monthly payments to the agency. The mill's loan carries no interest.

The mill has no special guidelines for judging to whom it gives a trucking loan. Generally the company must know the man who makes application; it also considers the amount of cane he has, how reliable a worker he is, and how much cane he will haul. A prospective trucker must have contracted in advance to haul at least 1,700 tons of cane (about 190 trips).

Running costs of a truck may be debited at specified service stations. The mill then pays the garages and deducts a like amount from the trucker's account. But this is not so much a service to the truckers as a practice insisted upon by the garages.

In theory transport prices are set by the mill and vary by distance. The longest trip should cost $2.00, but Los Boquerones suppliers usually pay a minimum of $2.20 to the nearest mill, and the price often is above this. The mill calculates that transport prices are too high and that with a reduction in such costs more would be left for the *colono*. Accordingly, it points out that the trucking system is a free market - each producer engages his own transporter and has the right to bargain with and change truckers at the end of the harvesting season. But, aside from the fact that the mill administers prices, few, if any, *colonos* take competitive bids for hauling from their friends. Furthermore, since the truckers are themselves *colonos*, a lowering of transport costs will only shift the distribution of the rewards in the countryside, not the division between the mill and suppliers.

The mill issues each transporter two entry tickets per day. The trucker is then held responsible for contacting his suppliers and arranging that sufficient cane be ready for hauling. It is questionable whether this pattern of 'farming out' the scheduling leads to the most efficient cutting of the cane in terms of optimizing sugar content and the use of labour; it does, however, reduce the mill's administrative costs.

Since a single haul takes between five and six hours, including loading in the field, driving, waiting several hours at the mill, and mechanical unloading at the grinder, many truckers find it necessary to employ a part-time helper. Both grower and trucker, thus, must be willing to stretch their labour time in order that the factory should not be idled. The machinery takes precedence over labour, the reverse of the relation between the peasant and his machete.

Upon arriving at the mill, the driver takes a 'sample' or selection of cane stalks from the truck, which is then tested in the laboratory to determine the sugar content of the load. According to the cane's quality, one of three prices is calculated per ton, $4.50, $5.50, or $6.50, although several factors eventually may change these prices slightly. Then the truck is weighed and unloaded. Based on the weight and average sugar content of his load, the *colono* receives a credit on his account. Sometimes, if a producer knows that his cane quality is poor, he may pay his trucker on the side to select a good 'sample'. Occasionally, if a man is in need of cash he may arrange to send a truckload of cane to the other mill or under a different name, thus avoiding immediate payment of his debt. Similarly, in the off-season cane futures sometimes are sold to truckers. In return for a discounted amount, the *colono* agrees to give a truckload of cane during *zafra* to his transporter, who can enter it to the credit of his own account.

On Saturdays during harvesting season, payments are made at the mill to both truckers and *colonos.* The trucker receives payment for his tonnage and distance less costs which the mill has paid. The producer receives a printout and the amount due to him after the various deductions. From this amount the *colono* pays his peons who harvested the cane during the week.

The acceptance of sugar cane

Both mills perceive a paradox in the way in which the peasants took up growing sugar cane. In the early 1960s the crop was readily accepted by the majority of *campesinos* to whom it was offered; yet, these same producers have been slow to accept techniques which might improve the productivity of the crop. Although this differential acceptance has baffled the mills, the reasons are understandable. The corporations' confusion stems from assuming that since the peasants took to growing sugar cane they must be using the same principles of production as the mills.

We may refer again to the two related forces of change in the rural area: factor push and market pull. Even discounting the effect of the sugar cane, *per capita* natural resources have been diminishing in the countryside, with the result that the productive base of the peasantry has been narrowing. As they were forced out of some of their traditional productive activities, the peasants turned to the growing of sugar cane, a seeming solution.

But the other force for change has been equally if not more pervasive. As one older man said, 'there is an enthusiasm of the people for money'. Money earned is not stored and hoarded, nor is it used as a fund for investment; it is expended on food and market goods, luxuries. Sugar cane provided the long-awaited opportunity for participating in 'civilization'. Old desires have been awakened, new needs generated. Suddenly the people have had the wherewithal to possess some of the goods which seem to define the powerful. 'In the old days there were no radios or dances with bands, and people made their own clothes and hats. Today people want money to buy these things.'

Initially sugar cane appeared to be an extra, a *lagniappe.* With ease the peasants were able to assimilate the cash crop into their traditional agricultural cycle. Their only added costs were seeding, weeding in the second year, and harvesting; but the mills even paid cash for these tasks. The risk of planting cane seemed slight since the peasants themselves invested no money or unremunerated time, and since the mills only collected against the cane harvest. Also, unlike rice, loss of the cane crop would not lead to household failure.

In this initial context the peasants had no reason to adopt more capital intensive, costlier techniques, based on balancing a flow of costs and revenues. Sugar cane added wealth to the subsistence system, while the subsistence cycle permitted and supported the introduction of the new crop.

But this short-run payoff has, of course, brought a longer-term cost: the increased depletion of the natural resources. The transition from rice to sugar cane is not only irreversible but cumulative, and by 1966–7 these resources costs were becoming evident and being discussed by the peasants. The new effects in turn are leading to an initial acceptance of the use of fertilizers and ploughs. But this second level of acceptance, unlike the first, is being taken up out of necessity.

As a result, the peasants now hold a bifurcated, almost contradictory, view of the mills. Some avow that it was only the mills which 'saved' the situation. 'The government did nothing, but the mills provided loans, and were it not for the sugar cane everyone would be starving.' In contrast, others claim that the mills caused the situation, that all their work now is for the mills, and that increasingly they are being governed by these firms. Rice, they say, is preferable to sugar cane.

Sugar-cane value and profit

'Progress', however, seems inevitable and perhaps this transition to a new agricultural crop represents a financial benefit. Despite the *campesinos*' misgivings, cane producers may achieve a higher standard of living than traditional farmers.

To calculate the value of sugar cane I shall use figures comparable to those enumerated for the rice and maize. Although the study of the sample households included questions about sugar-cane production, the resultant numbers cannot be used directly for a comparison. Three problems obtrude. First, sugar-cane production, unlike the subsistence crops, involves 'sunken costs', investments of labour and cash over a period of years before payoff. Even assuming no dating problem in the evaluation of expended funds, it remains impossible to calculate the value of a crop using only the figures from one year. What a peasant receives in the harvesting season is the net cash balance between his current harvest and his costs of the prior year, which in turn have been incurred partly for future years. Furthermore, each *colono* has fields at different stages of production and receives each year not a profit per field but a net cash flow balance for all the fields. Moreover, even if the peasants themselves conceptually separated the past costs for each of their fields, which they do not, it would remain difficult to calculate profits since the actual harvests of the different fields are cut and sent together; and the mills themselves do not try to distinguish old cane from new cane, field from field, in the printout which they give their producers.

A second problem arises because, of the fifteen persons in the sample, four do not grow cane, while five had in 1966 either a non-resulting or non-paying harvest. Only six actually harvested their cane and received a cash balance.

But even if these two problems could be surmounted, there would remain a third, more intractable, one. Since the peasants plant the sugar cane in their rice and maize, how are costs to be allocated among the crops? Although the sugar cane is added to the rice and maize midway through their cycle, its own costs do not begin at this stage, as the initial land preparation benefits all three seedings. The fact that sugar cane can so easily be added to rice and maize is a primary reason why it has been adopted, but its true costs - aside from resource devastation - are larger than they might initially appear to be. Therefore, to have an accurate comparison, sugar cane must be treated as if it were raised alone.

To construct the figures for one hectare of sugar cane, I have drawn on the available sample data, on observations and discussion, and on information from the mills. The socially expected revenues and costs for a sugar-cane plot are presented in Table 9, while a flow chart for each ton of sugar cane is displayed in Figure 2.

For each entry in Table 9 I have again been 'conservative', but the word here means the reverse of that which it denotes in rice and maize. Conservative sugar-cane figures minimize costs and maximize revenues; they favour sugar cane.

TABLE 9 Sugar cane: expected costs and revenues for one hectare

		Low	High
Harvest sizes in tons			
First harvest		30	40
Second harvest		20	30
Third harvest		10	25
Fourth harvest		—	8
Total		60	103
Labour days			
Cut forest		10	15
Make firebreak		2	4
Clean after fire		5	3
Seeding		12	11
First year weeding		20	17
Second year weeding (before first harvest)		20	20
Third year weeding (after first harvest)		10	7
Total		79	77
Harvesting @ 1 ton/day		60	103
Total		139	180
Revenue			
Gross/ton	$6.45		
Less transport	−2.25		
Net/ton	$4.20		
Revenue per hectare		$252.00	$432.60
Less seed cost		−15.00	−15.00
Net revenue		$237.00	$417.60
Profits			
Revenue/hectare		$237.00	$417.60
Less labour @ $1.25/day		−173.75	−225.00
Profit/hectare		$63.25	$192.60
Profit/ton		$1.05	$1.87
Revenue/labour day		$1.70	$2.32

The figures once more are presented as a range. On the low side I assume three harvests from a field, while on the high end I presume four. The labour days for land preparation are the same as for the rice and maize. The figures for the first-year weeding are conservative in that when sugar cane actually is planted with rice and maize, it usually receives one weeding with these crops and a second, briefer one, as the field is cleared of the dead rice and maize stalks after their harvest; effectively sugar cane receives two weedings in the first year rather than only the one which I include. For harvest-

FIGURE 2 Flow chart of revenues in the production of one ton of sugar cane

One ton of raw cane yields approximately 180 pounds of refined sugar; one pound of sugar retails in Panama for $0.11 and is sold by the mill for $0.096.

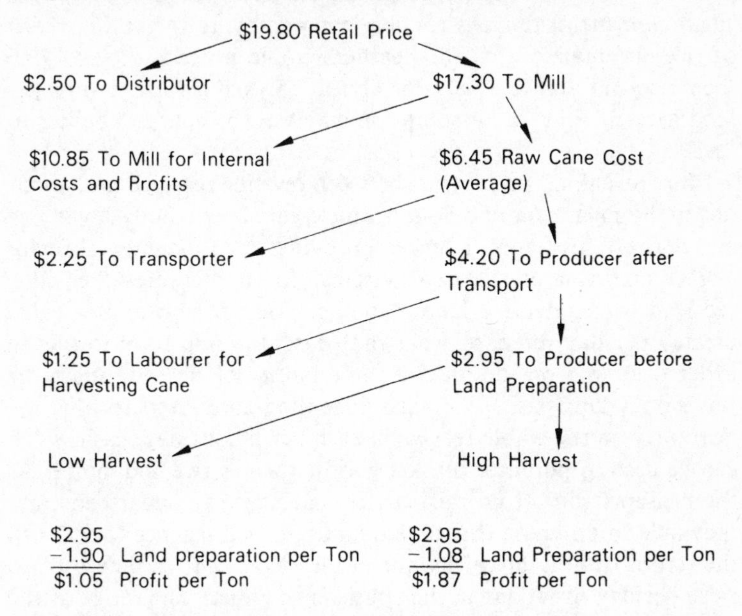

ing I assume the maximum output per day, though peons often do not cut and load a ton of cane in a working day.

Gross revenue per ton varies from $4.50 to $6.50 per ton of raw cane. Although one mill estimates that on average it pays $6.45 per ton, its actual figures ($980,000 gross paid ÷ 153,963 tons received = $6.36 per ton) do not support this claim. None the less, to be conservative I have used the mill's $6.45 figure. Transport costs vary between $2.00 and $2.50 per ton. The initial seed usually is purchased at a price of $1.00 per 100 stalks of cane.

That the final results are 'fair' to sugar cane is suggested by the fact that the midpoint of my profit per ton range is $1.46. One of the mills estimates that its *colonos*, most of whom live closer to the mill than the people of Los Boquerones and thus have lower transport costs, make about $1.50 per ton of raw cane.

The resultant calculations in Table 9 may be compared with those in Table 5, which are for rice and maize in one hectare. If the sugar-cane profits are annualized, without discounting, the profit range is $15.81 to $38.52 per hectare per year. This profit range falls far below either the sample average or the socially expected figures for the buying profit of rice and maize, that is what the peasants make as subsistence producers. On the other hand, the sugar-cane profit range *is* roughly comparable to the selling profit of rice and maize, what the peasants would make if they sold all their rice and maize. The crops appear to yield similar returns per land area, if sold - an assumption which is, of course, counter to fact.

More revealing is a contrast between revenue received per labour day in the sugar cane and dollar value created per labour day in rice and maize. This figure, I suggest, provides the fundamental financial comparison of the two systems, for it measures the value received for equivalent labour inputs. For rice and maize, value created per day refers to what the peasant would have to pay in order to obtain on the market the quantity of rice and maize he raises on a daily basis. For sugar cane the figure refers to what the peasantry on the whole receive in cash per labour day, cash which can be used to purchase market goods, such as rice and maize. As the peasants shift from raising rice and maize to sugar cane, are they able to purchase the same quantity of subsistence foods with the return from a day of labour in the sugar cane as they used to raise per day of labour in the subsistence crops? The range of the sugar cane figures is $1.70 to $2.32, with a midpoint of $2.01. In rice and maize the range of the socially expected figures is $2.06 to $3.37, with a midpoint of $2.72. As a crude measure the two midpoints may be compared; switching from rice and maize to sugar cane produces a value loss of over 26 per cent for each day of labour. Even the current sample average of $1.86 created by a day's labour in rice and maize, a figure which itself has been inflected downwards by resource diminution and planting of sugar cane in the rice and maize fields, is almost equivalent to the expected cane value. And it should be recalled that my conservative accounting is designed to favour sugar cane.

But there is a final and interesting dimension of these sugar-cane figures which should not go unnoticed. The mills advance to the *colonos* $1.50 for each day of labour they invest; *colonos*, however, usually pay their peons only $1.25 for the day, pocketing the differential. I have used the latter, actual, figure in the profit calculation for sugar cane. If, however, a *colono* actually paid the legal minimum of $2.00 per day he probably would incur a loss on his sugar cane, while if he paid the $1.50 a day figure, he would scarcely clear a profit.

The actual $1.25 payment may be contrasted also to the calculation of revenue per labour day in sugar cane ($1.70–$2.32) which is simply the net revenue received for the sugar cane divided by the total number of labour days. Comparison of the 'salary' and revenue per day figures shows that the *colono* eventually receives more than he pays his labourers. Without working he is able to keep a portion of the value created by the labourer. As the people explicitly point out, if they hire eight peons, they will pay each $1.25 and themselves $3.50 for the day; and the *colono* himself usually does not perform manual labour. Put differently, the more the *colono* employs others, the more he receives for his own labour. Marx identified this sort of 'exchange' as being fundamental to capitalism; others might argue that the differential is the reward, the profit, accruing to the entrepreneur for his organizational abilities. Either way, it is apparent that the importation of outside 'cheap' labour from the Santiago labour market is an essential ingredient in the system. For the profitable production of sugar cane by *campesinos* there must be a 'reserve sack of potatoes', to mix two well-known metaphors. Viewed from within the community, a peasant who works for other *colonos* is 'exploited' by his neighbours but exploits them in return. Viewed from without, exploitation is 'exported' to non-cane producing, more distant, villages. Is it tautologous to say that value flows up the economic scale?

The command of finance

Even if switching from rice and maize to sugar cane were a trade-off, even if the peasant received equivalent returns from both production processes, less visible aspects of the transition render its value equivocal, for this is not the type of change in which one car model is replaced by another in the production line; control of the line itself is shifting.

We may consider this transition from three perspectives. In terms of certain current economic categories, the shift from subsistence to cash cropping appears to be a gain. The agriculturalist who previously could grow only for home consumption now has the choice, the new freedom, of raising sugar cane and selling it for cash, which in turn can be used to purchase goods - amongst which the agriculturalist can choose freely - on the market. Further, the farmer can increase production up to the point at which he has available land, not just forest; his goals are expandable. And the entire venture may be undertaken at no monetary cost (though at the risk of indebtedness) since the mills provide interest-free loans.

Even within this framework, however, certain facts change the bright appearance of the picture. There are only two mills, not an infinity of purchasers, and the mills are not in competition. Conversely, the mills themselves face an almost perfect market of sellers, and they are able to 'lock in' these suppliers through the loans, which the suppliers are unable to repay except by selling to their creditor. The *ampesinos* are caught within a monopsonistic situation, an imperfect market.

But the new arrangement might be seen differently. The mills provide fertilizer, weed killer, and seed to the peasants on credit. Both firms construct roads and sometimes settle disputes, and both regulate the 'independent' trucking system. Wages and cane prices are set legally, but both mills have or had direct links with the political order. The mills, in short, assume social, political and juridical in addition to economic functions. Such an asymmetric, multistranded exchange between members of different 'social classes' might be categorized as a patron-client relationship; and it could be posited that the template, the model, for this bond was provided by the prior relationship between the peasants and cattle owners. If this earlier situation were viewed as being 'feudalistic', then the move to peasant cash cropping might be seen not as a transformation but as an extension and vestige of feudalism. At best the change in crops draws more money into the rural economy; at worst it is an historical regression.

Yet, this construction, too, omits important aspects of the new context, for aside from whether the prior situation actually was feudalistic, the new transaction between the peasants and the mills certainly is not, since it merely composes the end point of a series of exchanges which are linked to the international sugar market and consumption in the capitalistic world. The exchange would not be maintained were it not integrally linked to world capitalism.

Fundamentally, the *campesinos* are 'selling' not a crop but their very resources: the land, to which they have tenuous rights, and their labour. Further, since the planting of sugar cane has dramatic, irreversible, ecological effects, this is a cumulative process. To stay even, more cane has constantly to be planted. The market crop, by a perverse sort of Gresham's Law, drives out the subsistence cultivation. But as land and labour are increasingly sucked into, gobbled up by, the exchange, and diverted from subsistence, peasant control of these factors is lost. Their means of production are separated from them as the *campesinos* are brought into the national economy.

This control over sugar-cane production is achieved by the mills through their 'command of finance' (Robinson and Eatwell, 1973: 30). An original holding of capital funds is prior to the further accumulation of capital. With their financial command the mills are able to loan money for all stages of crop production. Effectively, they pay for and 'own' the seed, the labour invested and the product itself. At harvest time the peasant must turn over the product to liquidate these advances. Whether the final sums which the *colonos* receive are to be termed profits or wages - and even the mills cannot separate *colono* 'work time' from *colono* 'entrepreneur time' - is but a matter of terminology. Effectively all that the peasant as 'owner' does is subcontract his labour, and this, of course, saves the mills a certain amount of administrative cost.

From this perspective the fact that the loans are interest free is not an economic anomaly, an expression of goodwill by the mill owners, or a feudalistic vestige; nor are the loans a transaction between two capitalistic enterprises. Rather, the loans are an input by the mills in the form of a wage advance and an investment in some of the tools of production; indeed, no factory owner charges his workers for the use of his machines. If the loans were true loans, then the peasants could bargain between the mills for better raw cane prices.

None the less, the mills have limited their investment and financial command to certain facets of the production process. Control of peasant production is not achieved through actual land ownership, yet a further factor differentiating the new relationship from any supposed feudalism. By not purchasing land - an explicit part of their current strategy - the mills minimize their fixed investment or constant capital costs. Management of the raw cane supply is achieved solely through a short-term or working capital investment; their financial input is primarily in the form of wages or 'variable' costs.

From the peasants' standpoint there are both continuities and changes in this transition. Control over a monetary fund is not crucial either in subsistence or cash cropping. In subsistence, financial inputs are minimized; in cash cropping the investment fund is provided by outsiders. With the shift from rice to sugar cane the peasant's instruments of production - his machete and digging stick - change little, although in the long run they, too, will be transformed. At first, the peasant is able to preserve the illusion of being an independent farmer, of producing for himself, although actually he is labouring for others and is alienated from his product.

As the peasant produces more for others, however, so also must he purchase more of his needed goods. The peasants become increasingly susceptible to 'what the market offers', and this change is marked by a narrowing in the meaning of the word subsistence. It is Janus-faced no more. As the peasant plants sugar cane, subsistence no longer refers to producing for consumption, it becomes only the standard of living. And this standard, which previously was defined in relation to others in the countryside and controlled by the individual labourer, now is defined in relation to other strata in Panamanian society and controlled by market prices in relation to the wage in sugar cane. The actual objects consumed as necessaries may remain the same, but the system surrounding them - the way of obtaining and evaluating them - changes. In this fashion a peasantry may become 'impoverished' by moving into capitalism. The 'underdevelopment' of a rural area may develop from the advent of capitalism itself.

But all this returns us to the problem of distribution. The productive system in Los Boquerones is changing not just as a result of an alteration in productive methods but in response to distribution. Who has what is an integral part of the transformation. The production switchover could have happened only in the context of a prior distribution - the sugar-cane mills with their command of finance, the *colonos* with their command of land, and the peasant workers with their command of labour. 'Class' relations enter into the new form of production in the countryside. The eventual distribution of goods between these classes, then, is not a 'later' result of exchange and the market but a function of the 'prior' productive system.

But overall perhaps there was a certain inevitability to this transformation, for in so far as the rural subsistence economy was formed on the margins of a capitalist system, it could be maintained only as long as the encompassing economy had no interest in the outlying areas. Once this condition changed, once the driving wedge of a financial fund separated worker from product, the demise of subsistence was written in its genesis.

7 Horizons and reflections

The 'mixed' economy of Los Boquerones cannot be expected to endure; its current status surely is not permanent. Foreseeing the future, however, is riskier for the anthropologist than is planting subsistence crops for the rural dweller. But without prognosticating we can survey the problems of cash cropping and try to assess the potential of sugar cane for both Panama in general and Los Boquerones in particular.

Assessment implies the use of guidelines, and the outsider has neither the right to state what a country's objectives ought to be nor to specify how such goals ought to be decided upon. Therefore, I shall assume only that some national objectives are to increase production, surplus, and wages. In terms of these goals, what are the implications of abandoning the subsistence crops and of seeding sugar cane on an ever-increasing scale in Los Boquerones?

At the national level a major difference between sugar cane and rice is that the former is in part an export crop. Exporting stimulates the economy, for the revenue received from international sales can lead to a rise in internal demand, a demand which is unmatched by domestic supply. If the proceeds from exports are spent in the home market, exporting boosts the economy. From a national standpoint the raising of sugar cane has certain advantages over rice, at least in theory.

In practice, the crop switch-over raises as many new domestic issues as it solves internationally. First, there is the problem concerning the magnitude of the surplus. Measured in labour units how does the overall surplus (corporate and producer dividends plus new investment) created by sugar cane compare with that yielded by rice and maize? The calculations, of course, are complicated. More to the point would be a comparison between the surplus which is created in the sugar cane and the surplus which would be yielded in rice and maize if an equal amount of equipment were devoted to them.

By raising sugar cane the Panamanians also are diverting their resources from domestic agricultural products to exports. The 'value' of doing so is much dependent upon the vagaries of international exchange rates and world sugar prices. Will the revenue received from the sugar cane purchase as much rice and maize on the world market as is lost by not investing equivalent resources in these crops? In contrast to the present direction, a case can be put forth that a country ought to concentrate first on the production of its own primary crops. Given that rice and maize enter into the purchases of nearly all Panamanians, then an increase in the productivity of these crops ultimately will lower all costs and hence increase the national surplus. Not without reason did Ricardo and Marx emphasize the critical role of agriculture in an economy.

But how does sugar cane fit into this context? Since the product, sugar, is both exported and consumed domestically, is it an essential or non-essential commodity?

Although in limpid economy theory commodities can be sorted into categories, the realities of life which the anthropologist encounters are messier. Actually, the ultimate impact of an export crop like sugar may depend not so much upon whether it is also consumed by domestic workers as on what happens to the surplus it yields. But this raises a new series of questions, queries which are sited on the borderline between economics and anthropology.

Who is to control the surplus created by raising sugar cane? Assuming that the sugar cane surplus is equivalent to new (not replacement) investment in equipment plus dividends, then who determines this division between accumulation and consumption except the owner-managers of the sugar-cane firms? Who determines whether the major part of the surplus is to be spent on luxury, non-basic commodities, which are probably imported - thus offsetting the value of exporting sugar cane - or on investment for greater internal growth? The question is particularly crucial in the case of Panama, an international transit area with easy access to world products. Since there is little evidence that the country's elite 'buys domestic', the national value of raising sugar cane as opposed to rice and maize is questionable.

Then, there is the surplus question concerning the division between profits and wages. Who again determines this division except the mills? Overall, raising sugar cane requires a greater investment of resources per land area and represents the use of more productive methods than rice and maize. But the peasants are not participating in the fruits of this higher productivity and greater surplus, for at best their current standard of living is equal to their older one. In real wages, only under optimum conditions can they attain the same subsistence level as previously. Clearly, the owner-managers are receiving the major part of the new surplus which is being created in the sugar cane. For the peasants as workers to benefit from increased productivity, the ratio of profits to wages (the 'rate of exploitation') must remain constant or fall as output rises. But given their lack of trade unions, indeed given the vehemence with which the mills have fought syndicates, it seems doubtful that the *campesinos* will experience a marked rise in their standard of living.

We may carry this a step further. From the economic standpoint of 'realizing profits' or creating 'effective demand' there is little motivation for the mills to increase the peasants' return. Because further sugar output will be sold on the world market, the mills do not have to concern themselves with whether the peasants have sufficient wages to purchase their product. Only if the sugar were sold exclusively in Panama would the mills have to make certain that enough cash flowed back to the labourers so that demand could be kept up. It follows that the motivations of the mills for increasing the peasants' share of the product are solely political and social: as one of the mills pointed out, they have a conscience and are concerned about the threat of another Cuba.

If we shift our sights from the national to the local level of Los Boquerones, the 'value' of sugar-cane production appears equally nugatory. From the standpoint of economic development in Los Boquerones and given the current conditions, two directions can be envisaged. First, agricultural productivity in general might be raised. It is true that the mills have been trying to increase the productivity of sugar cane in particular but with the principal purpose of enlarging their own profits. Regardless, in the present situation there are some obvious, severe limitations to increasing productivity. Present plot sizes militate against the economic use of large-scale agricultural equipment. Increased agricultural productivity also will require more 'infra-structural' investment, a diversion of resources into such things as better roads and *campesino* education; however, both the national government and the USA through its AID programme have been slow to make such commitments. To attain a true increased agricultural productivity, a concerted effort and investment at the local level will have to be made.

A second direction for economic development would be to ensure that more productive use is made of the peasants' current surplus. To cite one example, it is not inconceivable that in their surplus time the countrymen could construct equipment for processing rice, such as making cement slabs on which to dry it. Eventually they might be able to control the whole of rice processing, instead of selling the raw product to millers. The increased surplus could then be used by the people themselves to raise the productivity of the crop. This second direction is different from the first in that it calls upon using the traditional system to build upon itself, to utilize its current surplus in a productive, cumulative fashion. Again, however, this path of development would require an outside commitment of resources - even protection - by a national or supra-national agency. It would also require the 'gearing down' of some of the equipment used in industrialized societies to fit the conditions of the countryside. The most 'capital intensive' machines are probably not appropriate in situations of 'gearing up' a traditional economy.

But all such sketchy, economic speculations pale beside the more massive and central issue of organizational change. If my analysis of the social conditions of the rural economy is correct, then any form of development, under whatever political guise, will and should require the elaboration of new social organizations in the countryside. The issue here is not only that of increasing economic efficiency, of surmounting the fractionary effect of household organization, but of social and political participation and control. The most radical change will come only with the development of 'median' organizations, organizations above and beyond the household level yet not so large as to be removed from peasant control. But let me be clear. In the context of the Panamanian countryside certain older and single-factor theories of development do not capture the essence of the problem, theories such as those emphasizing the need to change the 'people's values', raise achievement motivation, hold educational classes, demonstrate new techniques on experimental plots, enact land reforms, extend rural credit, or make available high-yielding seeds. The critical economic question is not in itself economic but social: control or partial control of the productive system. To be sure, agricultural innovations are important, but the thrust of my argument is that few, if any, social-political-economic organizations mediate between the peasant and the larger Panamanian society. Injected into this vacuum, purely 'economic' aid will do little more than polarize and accentuate the already existing division between the peasants and the national society.

To propose that new forms of organization are needed is one thing, to suggest their form another. Unlike, for example, New Guinea with its patterns of 'Big-men', or parts of Indonesia which had traditional collective organizations that could be adapted to new functions (Geertz, 1963), the Panamanian peasants have no customary patterns of leadership or communal groups that can serve as the conduits or models for development; there can be little continuity between the old and the new. Who, then, is to specify what the new organizational forms should be? The impetus for developing the new forms will have to come from without and the organizations themselves will need to be supported from without, but if the people of the countryside are to benefit, they themselves must help develop the new collectives. Perhaps the saddest legacy of the past 400 years is that even the organizational forms of 'modernization', under whatever national form of government, will initially have to be implanted on the countryside.

I began by asking how productivity and surplus might be increased in the countryside, but this query has led to the far more perplexing question concerning social forms and changes, or to the problem of control of resources and distribution. This is perhaps the ultimate conundrum. It is hard to see how much 'economic progress' can be made in the countryside without massive outside intervention, yet such an intrusion must respond to, indeed receive an impulse from, the countryside itself. How can the rural area stimulate outside forces to enter the countryside in order to benefit and to be under the control of the peasants themselves?

The Panamanian situation is not unique, however, and in many respects I have provided a case study of a more general process, one example of the inter-digitation between a subsistence system and capitalism. To see how this example relates to others we may take a slightly more general perspective.

Consider the relation between the two contrasting economies in terms of production. In simplest terms subsistence as a productive system consists of combining land, seed and labour to yield an output:

Land + seed + labour ⟶ Crops

In Los Boquerones the intrusion of the outside system had two general effects on the existing system of production. At first, labour was only mildly diverted from the subsistence crops; most of the work put into cash cropping came from the surplus time which was created by the subsistence economy itself. On the other hand, capitalism did have a critical impact on land use; the competition between the two systems occurred over the use of natural resources. Both economies were competing for the same resource niche.

But what if an intruding system does not divert the land, or more broadly the resources, which are being used by the indigenous system? What would be the consequence? The impinging system may have found or be able to make use of a resource not utilized by the existing economy, such as mineral deposits. Or, through the use of equipment, the outside system may be able to make use of otherwise untillable land. In such instances the new system does not undermine the old one from a resource standpoint; the two can co-exist. The crucial variable, however, becomes labour. The effect which the outside system has depends on whether or not and in what fashion it draws upon the indigenous labour force. Indeed, we may expand this second form of economic inter-digitation to include cases in which the outside system is not even spatially juxtaposed to the existing economy, a typical instance being the use of migrant labourers, workers who leave home to earn cash on a seasonal or longer-term basis. Well known throughout the world, this form of economic intrusion is exemplified in Panama by the labourers from distant areas who come to Los Boquerones and to the mills to earn money working in the sugar cane, and by the peasants elsewhere in the country who work part-time on the coffee and banana plantations. In such instances of labour use the subsistence or home economy through use of its surplus may actually 'support' capitalism; the per day value equivalent the labourer receives in cash wages may be less than what he creates each day in his own agricultural work. The subsistence system may provide the infra-structure for the capitalistic venture.

This second type of situation, which is sometimes said to be a function of 'sub-employment' in the rural sector, can have its material benefits for the subsistence farmers. Much depends, however, on when their labour is used and on how their cash wages are spent. Sometimes the earnings of the labourers are injected into the productive system of the native economy; more commonly, the wages are used to buy consumer goods, which in turn helps the outside producers to 'realize' their own profits. When the wages are used to keep up the level of effective demand in the national economy, they have little or no effect on the subsistence economy, other than to provide temporary material benefits. Only when the cash is invested directly or indirectly (through equipment purchases) in the local productive system does it lead to increased output and accumulation in the native economy. To have such an effect a mechanism must exist, a structure created, so that the workers can invest for their own benefit the cash they have earned in the surplus time they originally created.

On the other hand, it is not to the benefit of the impinging economy to divert so much labour from the local system that the latter is undermined. If the labourers are drawn off the land entirely, or even during the crucial periods of the local productive cycle, the result is under-utilization of natural resources in the countryside. More important, the workers themselves will need to be paid a higher wage, to receive a larger share of the output, in order to preserve their pre-existing standard of living. Such 'competitive' use of manpower seems rare in the Third World today and probably occurs only where the national economy is expanding rapidly and is experiencing a labour deficit relative to other inputs.

Lastly, in the third case, the outside system may divert both natural resources and labour from the native system. Although in Los Boquerones the initial clash between the two systems occurred over land, increased planting of sugar cane will divert more and more local labour to the crop. The result will be the gradual but complete undermining of the subsistence system and the conversion of the peasants into a rural proletariat.

At the outset I raised the hope that anthropological economics might be revivified, that it might rise above its petty battles and short-sighted theories to take its rightful place in the social sciences. My effort here constitutes, I hope, only the beginning in the development of a serviceable and sensible inventory of ideas. Above all, we - anthropologists - must first but finally consign to the dustbin such shop-worn concepts as capital, not for ideological reasons but because capital is such a muddled idea, at least as used by anthropologists. By capital do we mean, in the Marxian sense, a certain type of relationship between owner and worker? Alternatively, does capital refer to a monetary fund; but how then does it differ from savings? Or, is capital a fund of money actually advanced to workers and expended on raw materials and equipment? In this last case, is there a difference between 'working capital' and 'permanent capital'? What time period (a year?) should be used to draw the dividing line between the two? Further, in exactly what sense is physical equipment - machines and tools - capital? Capital, in short, is a synthetic idea, a cultural concept used by some economists and the people of certain societies as a means of grasping experience. At a minimum, it can be used only in those contexts where different qualities and objects can be reduced to a common denominator or currency, but even then one wonders whether it is a term of analysis or of justification for the reward rentiers receive.

Despite these limitations anthropologists have gaily forged ahead using terms, like capital, in rather loose ways. A case in point is Salisbury's (1962) celebrated study of the Siane of New Guinea. As Salisbury documents, the Siane some time after 1933 were introduced to the steel axe, a tool which quickly came to replace their stone implement. Salisbury has much of interest to say about this change, but his theoretical ammunition is not satisfactory. Although Salisbury is quite aware that 'capital' conveys different meanings even to neoclassical economists, his main concern is to arrive at a definition of the term which can be adapted to the pragmatics of Siane life. Thus, for Salisbury, Siane capital includes stocks of 'valuables' (1962: 153), some finished goods, fixed or durable production tools, certain consumer items such as houses and clothing, and 'liquid' or 'working' goods, such as raw materials and gardens in process. This is truly an heterogenous collection, all subsumed under one label. Even more, since the Siane do not possess a general currency which might be used to 'reduce' all these items to a standard unit, Salisbury is confronted with the problem of comparing this disparate collection. His solution, paradoxically, is to suggest that labour cost might be used as the standard for comparison; but even then Salisbury shrinks away from the radical direction of his thought, for he submerges labour under the concept of time and suggests that 'this index may only be useful for Siane' (1962: 144). In different terms, and somewhat more formally, we may say that Salisbury essentially wants to reduce all costs to dated labour; goods represent the expenditure of past labour. But exactly why Salisbury wishes to delimit capital to certain past labour expenditures, and not others, remains unclear.

In fact, we may outline the rudiments of the Siane transition from stone axes to steel axes in rather simple terms. This was a change in tool use which had two different consequences. First, in terms of producing the equipment, from the Siane standpoint steel axes take twice as long to 'make' as do stone axes, 12 days to 6 days (1962: 145). Steel tools represent a larger amount of accumulated labour than do the stone ones. On the other hand, a steel axe lasts twelve years, a stone axe but a year and a half. The net result is that on a yearly basis a steel axe represents one day of accumulated labour, while a stone axe represents four times this amount. In essence steel axes can be 'made' more quickly, and for this reason alone a greater surplus - to be utilized in various ways - was created by the change. Second, in use the steel axe is more productive than the stone one; with use of the steel axe output per labour input rises, such that the same labour input as previously yields a greater output, or a smaller labour input produces the same output as before. Numerically, for the Siane, the only work change was that the time for clearing and fencing a garden was reduced from 36 to 12 days. Again, the ultimate impact was to increase the size of the surplus, as would be expected with an increase in labour productivity. In the simplest of terms the overall yearly surplus was increased by 27 (3 + 24) days. Given this analysis, it is difficult to understand how Salisbury can claim that the change did not lead 'to any substitution of capital for labour in production' (1962: 150), unless there is a problem with his use of the term capital.

Of course, we can from here trace the effects which this equipment transition had on Siane use of 'leisure time' and then explore to what degree the new surplus was or was not reinvested and cumulated in the system. But this is as far as I wish to carry the example, for my point is that much of what Salisbury said was obscured and confused by his use of the concept, capital.

But Salisbury is not a unique offender, and in certain respects a much more difficult and complex example is provided by Epstein's (1968) analysis of the Tolai on the island of New Britain. Here, to be brief, Epstein appears to make the assumption that hoards and stocks of valuables accumulated by pre-contact 'Big-men' were equivalent to 'capital funds' accumulated by modern entrepreneurs (1968: 218-19). Since any agglomeration of funds or objects is 'capital', Epstein seems to feel justified in speaking of *Capitalism, Primitive and Modern.* Once again, the word 'capital' has made it difficult to grasp both the similarities and differences between two systems. Indeed, Epstein's description of the traditional Tolai is rather ambiguous. On the one hand, she states that 'The Tolai "big man"' like a true capitalist, invested his resources in order to increase his wealth' (1968: 28), but this is shortly qualified by 'There was no way open to a "big man" to convert at least some part of his tambu wealth into some other durable asset . . . . All he could do was to translate his economic achievements into prestige and political influence' (1968: 30). Further, though Epstein holds that the pre-contact economy was a form of 'primitive capitalism', she has some difficulty explaining why the economy was stagnant. She grants that the Tolai lived in an area of 'highly fertile' soils, but finally argues that 'there are no readily available minerals in the area' (1968: 32).

Clearly, in the Tolai case it is important to distinguish carefully between hoards or savings of wealth objects which are redistributed and exchanged, and funds or stocks which are reinvested in production and have a cumulative effect. One may call the hoards and the funds by any terms, but to label both 'capital' and then proceed to equate them is quite unjustified. Accumulation of wealth through exchange is not the same as accumulation through production, and both must be distinguished from the idea of productive investment or growth (Robinson, 1965a: 256-7).

In both the Tolai and Siane cases, analysis and understanding would have been enhanced by avoiding words like capital and by returning to fundamentals, to the very concepts I have used in the course of this book: concepts such as labour, tools, productivity, subsistence, surplus, savings, and reinvestment. I call upon my colleagues to begin again in order to develop a usable 'stock' of ideas. We must, as against the substantivists, focus our attention firmly on humans as opposed to the 'substantial' objects they shuffle about; and we must, as opposed to the formalists, shift our attention from exchange and consumption to the essential and primary act of production. Goods are created by human labour, and ultimately, I have argued, the concept of labour - of humans engaged in the process of production - can provide one foundation for a cross-cultural economics. How is labour conceived and organized in different societies? Tools, like modern machines, are equipment, and we need to look again at the man-tool relationship not to differentiate us from the apes, not to make inventories of material culture, nor because a 'materialist' approach is 'correct' *a priori*. Tools represent and embody past labour, and have an effect upon current labour's productivity. Under what conditions and how does labour get 'stored up' in tools? Why does man 'extend' himself through machines, and machines that make machines? Is, in a particular society, the stock of equipment being augmented or not, and what determines the equipment level? We need to confront again the nature of subsistence and of surplus and to make a direct attack on the problem of distribution in classless societies, for it is precisely here - the siting of distribution within production - that the anthropologist has a unique perspective to offer. I have proposed a cross-cultural, technical way of understanding subsistence and surplus; but to grasp the rationale of distribution we must consider its social setting, for it is the social relations of production which lie behind the observable economic facts. These are only the outlines of an anthropological economics, an economics which finally shifts the focus of study from objects which are exchanged to humans producing objects.

It is always easier, however, to proffer advice than to accept it. For years it has been said that among its virtues anthropology teaches 'cultural relativity' - that equally valid but different societies function throughout the world. The anthropologist's task is to understand, to explain, to translate these distant cultures to the people of his own. Of course, the question exists as to whether anyone can make this leap of understanding. How can one verify that the concepts of a foreign people are as one says except by ultimate reference to empirical facts, facts which are themselves framed by concepts?

But I would suggest that one final lesson of this book is rather the reverse of cultural relativity, for its leitmotiv has been that the economy of the rural dwellers of Panama is a creation of the same system of which we are a part. Diverse from us, at another level the *campesinos* are we. To speak, then, of 'changing' the countryside, of the 'special problems' of the rural area, is not only useless but false, for the countryside and the industrialized parts of the world are tied together by a skein of not entirely visible threads. To pretend that 'we' can help change peasant life, or in more liberal form help the peasants to change their own way of life, without also changing ourselves is a misconception. The evidence is there; what happens to the *campesinos* will very much be a function of what we do, and by this I do not mean the cauterized window-dressing changes so often promulgated by planners. Ultimately, the focus is on us, not the *campesinos*.

Yet, this takes us directly back to 'cultural relativity'. It may be that anthropology's unique lesson has been to provide a 'mirror for man', to see humans in the light of their full range of capacities. But this is too abstract. In place of offering a mirror on humankind in general, perhaps this book provides a partial self-reflection, for what we see in others is mostly an image of ourselves - what we want to see and what we are. Whether we like this reflection and what we do about it is a different issue. Here anthropology is mute, for it is but a hall of mirrors.

Epilogue: continuities and changes

Nearly ten years have passed since the time of fieldwork in Los Boquerones. But the changes wrought in this interval have been of even greater magnitude than those of the previous four hundred years. Most notably, the village economy has been totally transformed by the construction of a nearby government-owned sugarcane mill. The entire land area of the community is now devoted to raising sugar cane. In July 1974 and then again in March 1976 I was able to return briefly to Los Boquerones to witness some of the effects of this change.[1] In certain respects the past decade represents a new story; in other regards it is but an elaboration of old themes. For the people, the two crucial and new variables have been a subtle shift in power from private enterprise to the national government, and a dramatic rise in the world sugar price. Control still radiates from Panama City to the interior but it is filtered in a different way. A brief sketch of some of the recent changes may be of interest in so far as they represent a transition from 'capitalism' to 'state socialism', a process not unknown in other parts of the Third World.

In 1968 elections for president were held in Panama, and for the third time Arnulfo Arias, a 'charismatic' though somewhat ambiguous leader, was elected to high office. But within eleven days he was - again for the third time - deposed. The coup by the National Guard was led by Brigadier General Omar Torrijos Herrera, whose Revolutionary Government has remained in power ever since.

The Torrijos administration has initiated a number of changes throughout Panama, impulses which have had important effects internationally in relation to the Canal and domestically in terms of economic development. Behind both initiatives lies a new sense of nationalism and independence.

The new government has taken a direct interest in the rural area and the life of the *campesinos*. General Torrijos himself is a native of Veraguas, and he has recruited numerous people from the interior to hold important government offices, with the result that the power of the country's metropolitan elite has been challenged. Backed by military force, the Revolutionary Government has opened political and economic participation to a wider range of Panamanians.

One of the early acts of the government was to adopt officially a document entitled the Veraguas Plan (*Plan de Veraguas*), a guide for action in the interior which this writer helped develop with a group of Panamanians. The main purpose in publishing the Veraguas Plan was to emphasize publicly the diverse social and economic problems of the interior and to offer a variety of suggestions for development. The document underscored the importance of securing greater *campesino* participation within the country and provided evidence of some of the endemic agricultural problems. Among other suggestions it put forward were to develop food processing plants and co-operatives within the province.

Whatever the ultimate cause the Government wasted little time before turning its attention to Veraguas. Feasibility studies for building a new sugar-cane mill in the province were begun in 1969 and actual construction commenced in 1971. The mill, known as La Victoria, is located some six to seven miles north of Los Boquerones and is designed to draw on a circular supply area containing 10,000 hectares. In 1973 the first harvest of 246,199 tons was milled and this grew to 380,000 tons in 1974. But the mill's ultimate capacity is nearly twice this much.

Representing a $25,000,000 investment, La Victoria is both an integrated complex and larger than the two older refineries. The mill draws on nearby rivers for its own water and for field irrigation. Nearby lime deposits are used to provide fertilizer. The roadway network has been vastly expanded, some machine shops built, a small airstrip constructed and several planes for aerial spraying purchased. Most of the partially refined sugar is trucked to the port of Aguadulce, which itself has been improved, and shipped from there around the world. The mill maintains buses for transporting cane workers, and has dormitories and several restaurants. Increasingly, all its field operations are being mechanized.

The mill has the capacity to provide permanent employment for 500–700 workers, plus jobs for some 3,000 more during the harvest.

The cane-growing area itself includes 26 communities and 10,000 people.

Chartered originally as a government corporation, La Victoria, according to initial plans, was to be converted into a large production co-operative with collateral sections of savings and credit, consumption, housing, small industry, general services and transport. All workers associated with the mill would be eligible to join. By 1974 the mill legally was a 'pre-co-operative'; theoretically, a general assembly of *campesinos* elected and gave authority to an administrative council, which in turn held ultimate control over mill operations. The original aim of the government was to repay the outside financing within ten to fifteen years and then turn the entire mill over to the fully formed co-operative. One facet of this transition process was to include the training of local people in technical and leadership skills so that they might help to administer the mill at all levels.

Prospects were bright in July 1974. The breakeven point for profitable sugar production was calculated to be at a selling price of $0.12 per pound, but world sugar prices were on the increase from $0.15 to $0.20 to $0.315 per pound, a final price one person termed not only 'fabulous' but 'crazy'. Encouraged by success the government began planning for the construction of several more mills in other parts of the country. It appeared that the socio-economic aims of the venture would rapidly be brought to fruition.

By 1976, however, the picture had changed. La Victoria had expanded operations and made economic progress, but the promulgation of its social goals had slowed. Plans for converting the mill into a co-operative had been abandoned and many of its original more liberal administrators replaced. 'Social studies' of the people in the area had included little more than the collecting of statistics about type of housing, how the people disposed of their garbage, the incidence of diseases, salaries earned, and the composition of the population.

The advent of the mill has, of course, had a remarkable impact on Los Boquerones. The government corporation assumed all previous debts to the older mills which, incidentally, are still operating but drawing on other production areas. It provided loans for building new cinder block homes, and all the *campesinos* were moved to housing sites bordering the Inter-American Highway. (One plan, never put into effect, was to agglomerate all the affected communities into three or four residential areas.) An accord also was worked out with the agrarian reform agency concerning land title. Through the mill the peasants are to be enabled to purchase the land. Measurements were taken of the plots which residents had encircled, then all fences were torn down, access roads constructed, and the entire area - forest, savanna, crop land and pasture - was converted to the raising of sugar cane.

The villagers now have nothing to do with the farming of 'their' land, the entire area being under direct control of the mill. Labourers are brought in from all parts of the interior, and teams of men, accompanied by heavy equipment, work their way across the countryside for seeding, harvesting and other agricultural activities. Because all the old physical signposts have been obliterated, it is impossible for an individual now to specify exactly which is his plot or when his land is being worked.

These new methods, of course, have changed the older production system within Los Boquerones. In the first place, the natural wealth in the cane area has been undermined to the extent that the *campesinos* cannot depend on the countryside for any resources. To cite a few examples: most of the people have tried to keep small gardens about their homes, but air spraying of herbicides and pesticides has killed many of their seedings. With the extinction of the forest no firewood is available and everyone now must possess a small gas stove, purchasing the propane in Santiago. All houses are roofed with corrugated tin not only because grass for thatching has disappeared but also because tile cannot be used since it requires an underlying lattice of thick wood, which itself is not available.

The older crops can no longer be raised, and the technology as well as all the terms the people once used - *monte*, *rastrojo*, *casqueado* - are no longer employed. The large wooden mortars and pestles for shelling rice sometimes can be found - rotting behind people's homes. Most households still keep chickens about, but the feed has to be purchased. Hogs are not raised, although a few people have been able to keep cattle on land not suitable for raising sugar cane. Every form of labour organization has fallen into disuse, and most of the older specializations, such as making horse blankets from reeds, have vanished. One horse-driven cane press remains, and a few persons occasionally make fishnets and go fishing.

The *campesinos*, in short, are no longer peasants - however that term be defined. In the light of their sources of income it seems most appropriate to view them as rentiers and as a rural proletariat.

In return for the use of his land, a man receives from the mill at harvest time $70 for each hectare which he has planted in cane, regardless of land quality and actual yield. Output per land area has, on the average, risen dramatically and this payment represents about $1.00 per ton of raw cane after costs, a figure lower than the countrymen used to receive per ton. But in addition to the rise in output per land unit, the average household probably has now between ten and fifteen hectares planted in cane, and some have considerably more. One man, who used to work sporadically and possessed almost no cane in the 1960s, received in 1974 over $600 for use of his land. Another had a return of nearly $3,000.

In addition, most of the men are employed by the mill, which has promised work to all those living within the cane area. The form of work varies from loading trucks and performing agricultural labour to being a field foreman or driving heavy equipment. Salaries vary in accordance with the skill level. At a minimum a man can earn $18 a week, though many take in nearer $30 per week. The younger men who have learned how to drive field equipment can earn up to $11 a day, and during the harvesting season, by exerting himself, a man can earn $20 a day cutting cane. Although the work week is six days and the work day nearly eleven hours, nearly everyone avows that the physical labour itself is now easier.

The wages, of course, represent a tremendous increase over nine years ago, and thus economic welfare or the 'subsistence level' has risen dramatically in Los Boquerones. The wage increase was made possible by the fact that exported sugar yields a greater surplus due to the increase in the world price and enhanced productivity. Put differently, the cost of living has risen in other national economies, their surplus has declined, as rural Panamanians have experienced a rise in their standard of living.

In practical terms, more public services are available to the people. An aqueduct system was completed in the community in 1976, and for $1.00 a month each household now can have running water from an outside spigot. Even earlier, electricity from a power plant, unconnected to the mill, arrived, and by 1976 nearly all the homes directly at the highway were electrified. Television sets are becoming as commonplace as transistor radios once were. Through the mill the people have begun also to receive health and pension benefits, and far greater use of doctors is now made. Personal consumption has, of course, also risen. Since food costs have less than doubled, they now take up a smaller portion of the worker's salary. One example will suffice; store-bought furniture has almost completely replaced the crude wooden benches that once were used.

The effects of these changes on the social organization and culture of the village have perhaps been slower to reveal themselves. In that households used to be linked by small-scale loaning and labour exchange, 'communal integration' has broken down. On the other hand, a 'communal work ethic' has begun to appear, though in nascent form. Some of the men point out that in contrast to the old days it is important to show up on time for work not only to receive a salary but because others are depending upon one for the fulfilment of certain tasks. Previously, a person sought to live *sin compromisos* - without obligations - but now one has a daily *compromiso* at the mill.

Family organization within the household has thus far changed little. The male still holds ultimate authority over the group and, as before, links it to the productive sphere. Whereas it used to be said that a man, to be a respected person, had to go to the fields, now the people state that men go to the mill. 'Salaried worker' has shifted from being a derogatory expression to becoming an accepted way of life. The life of the female has certainly become easier. The woman of the house must arise slightly earlier than formerly in order to prepare breakfast and a lunch to be carried to the factory, but some women now go back to bed afterwards. More important, a female now has to undertake fewer food processing activities given the new dependence on market goods. A few of the women also have obtained jobs at the mill, while several of the younger ones have become or are training to be school teachers.

In the family sphere the most marked change appears to be a slight cleavage appearing between the generations. The better paying jobs which require more training and skill are occupied primarily by the younger men, and they in turn have a higher standard of living than their elders. Young males speak with greater assurance and knowledge about current happenings than their seniors. Some young people allowed that when I first stayed in the community the people lived 'rustically'. Thus far this generational change has not had a reverberating effect on the social organization, for the elderly never held control over their grown children nor over critical resources. To the contrary, the problems of the aged have eased slightly in that they now can depend on their children for more help.

The mill has not had a direct impact on *campesino* beliefs, though changes in this domain are certainly occurring. In overall terms an opposition between a religious and a national ideology appears to be emerging, though to what extent this is a conscious effort on the part of the government is unknown. The entire system of belief in the saints, with the many ramifying meanings it carries and functions it performs (Gudeman, 1976b), is slowly being eclipsed. At a practical level the mill does not allow time off for the many saints' days, although a few are still observed; instead a number of civil holidays honouring workers and important days in national history have been instituted. This is not to say that the customs of making vows to saints and holding processions in their honour, practices which had assumed great importance in the countryside, will disappear, but certainly their future is uncertain.

More profound and intriguing is to consider what will take the place of the religious ideology which for the people provided, however incompletely, a set of meanings and understandings about the world. Here a brief account of a recent happening is perhaps relevant, for the occurrence not only gives a suggestive glimpse at the future but provides one (rare) vindication of the anthropologist's analysis. In two prior publications (1976a, b) I put forward the view that for the *campesinos* the Christ figure represents not resurrection and a new life but persecution, suffering, weakness and resignation. I related this conception to the socio-political conditions of the interior and to the male role specifically. Much the same view can be found in Chapter 3 of this book, and it is precisely this analysis which has been confirmed, obliquely, by General Torrijos himself. As part of his revolutionary stance General Torrijos stated and publicized a number of maxims intended to rouse the populace. For example, at the mill the following slogan uttered by the General was painted on a wall:

On the feet or dead,
Never on the knees.

In 1976 I learned that another epigram had been offered by General

Torrijos, though after some opposition it ceased to receive official publicity. It is this slogan which is of special interest in light of my own analysis and the fact that Torrijos himself comes from Veraguas.

> The only thing that I have against Christ
> is that he died without fighting.

The passive religious image is not congruent with, does not lend itself to, the new conditions of Panama. Further, just as the subsistence system was buttressed by a coherent set of beliefs, a link is being formed between the local sugar-cane mill as an economic structure and a growing national ideology. The mill is more than a business enterprise, yet hardly likely to become a producer's cooperative. Currently it represents – it symbolizes – a set of ideas about national sovereignty and the ordered, military posture of the government. The mill stands for a 'revolutionary' break with the past, as will national control of the Canal. The link between ideology and economic patterning is no longer mediated and supported by religion; yet structural and ideological changes in the countryside still follow upon impulses from the transit area.

The villagers' thoughts about all these changes focus mostly on the practical. They are pleased by the easier work, not bothered by the longer hours, and happy with the material improvements. Many point to the expansion of opportunities which they see for their children. Their greatest concerns now are typical of the Western world. At times some of the men have been laid off work, and given their complete dependence on the mill this has proved to be a real hardship. In recent years inflation also has become a concern since it affects the people's buying power and standard of living. Few are certain about the future of their land ownership, but most are contented with the yearly payments they receive, and control of the land itself has ceased to be the critical factor in making a living.

But the new situation is not without its perils, and the ultimate direction of the changes is problematic. The mill has exhibited a conflict of purposes and a shift in its direction. For example, with reference to convincing the *campesinos* to switch from raising rice and maize to sugar cane, the mill maintains that it explained the change to the people and demonstrated to them what the economic advantages would be. Most of the people, it says, were swayed by the economic logic. (The peasants themselves did not perceive a choice.) On the other hand, when explaining why complete control over agricultural production was assumed, some mill administrators said (in 1974) that the *campesinos* are 'careless' and that the mill can accomplish the tasks more efficiently. In certain contexts, then, *campesinos* are viewed as rational decision-makers and in other situations as economically irrational. Alternatively, some might see this contradictory explanation as an after-the-fact justification for the power which the mill, backed by the military government, was able to assume.

Similarly, the mill has taken a narrow view of the meaning of social change. For the mill 'social advancement' appears to mean having better roads, electricity, water systems, better schools, and medical clinics. Few would disparage these economic advances, but these are not indices of organizational change in the countryside.

The mill itself, however, has been caught between the pressures of having to follow two, sometimes incompatible, paths. The original government investment was based in part on the idea of *campesino* betterment, of co-operatives, and of the devolution of economic and political power. Social not economic impulses lay behind the project. On the other hand, the mill must succeed economically, for a major portion of the investment funds has come through Wall Street. The quickest way to turn a profit and repay the loans was for the mill to centralize and assume complete control over production. From its inception the mill has embodied contradictory forces. At the moment, the balance has tipped in the direction of pursuing the 'more economic' path. It follows that in 1976 'social' change meant building better houses and improving health practices in the interior; the central social problems are perceived to be a high turnover of the work force and the inability to move the countrymen to more central living sites. Profit-making precepts are still having an impact on the countryside, at a step removed.

By 1976 some of the *campesinos* were becoming aware of their own 'loss of freedom'. As one put it, 'pressure was put on the old rich, but *campesinos* have not achieved true liberty'. The power of the mill and its link to the National Guard, the control the firm has over diverse segments of countryman life, the people's inability to form unions or to agitate for changes (except at the cost of losing one's job), the fact that the mill will not now become a co-operative and can at any time lower the payments made to the *campesino* - all are recognized facets of the new situation.

The older pattern of domination of the interior by the metropolitan area, with the accompanying notion that the rural dweller is 'backward', is yet in evidence. Ultimately, the choice has still to be made whether change in the countryside is to be instituted from the top down, the bottom up, or in combination. As their older economy, culture and social organization are destroyed, the question remains whether the countrymen will be allowed to take the initiative, the responsibility and the control for evolving their own response to raising sugar cane for the world market.

Notes

1 Anthropological economics and a small village

1 See Firth (1975: 40–1), Frankenberg (1967: 70–4), Godelier (1972: 295–6), and Salisbury (1962).

2 In a previous publication (1976a) I avoided use of the term 'peasant', citing the ethnographic fact that within Panama the people are termed *campesinos* or countrymen. But there is also a theoretical problem in that for some the word peasant may call forth an image of 'surplus' product being 'extracted' from cultivators. This has not traditionally been the case in Panama. Therefore, I shall use the term peasant only for the reader's convenience. For a more complete description of the community the reader may consult the above publication.

2 An economy evolves

1 The name Veragua was changed to Veraguas some time in the mid-1700s (Carles 1959: 272).

2 Rice was introduced by the Spanish probably in the early 1600s (Fuson 1958: 208).

3 'When in the progress of society, land of the second degree of fertility is taken into cultivation, rent immediately commences on that of the first quality, and the amount of that rent will depend on the difference in the quality of these two portions of land' (Ricardo, 1951: 70).

3 Household production: subsistence and surplus

1 A more complete analysis of the interplay between kinship and residence is contained in Gudeman (1976a).

2 Such a structural and economic view of 'spheres of exchange' may be contrasted to cultural (Bohannan, 1955) and 'rational man' (Ortiz, 1973) interpretations.

3 The people also avidly play the national lottery.

4 Using the same type of assumptions, Sen (1966: 426) states: 'labor is applied up to the point where its marginal product equals the "real cost of labor".'

5 For some of the many anthropological views, see Diamond (1974: 12), O'Laughlin (1975: 362), and Sahlins (1972: 91).

6 To recognize that there are marginal declines in land productivity, due either to bringing poorer quality land into production or using more intensively the same land, is not at all to offer a 'marginalist' analysis of the 'factors of production'.

7 If production days (P.D.) are increased by x, total production (T.P.) is

augmented by a factor of

$$\frac{\text{total production}}{\text{production days}} \text{ times } x.$$

The formula for rate of surplus is:

$$\text{Rate of surplus} = \frac{\text{Surplus}}{\text{Necessary maintenance}} = \frac{\text{T.P.} - \text{P.D.}\ \dfrac{\text{T.P.}}{365}}{\text{P.D.}\ \dfrac{\text{T.P.}}{365}}$$

The new formula would be:

$$\text{Rate of surplus} = \frac{\left(\text{T.P.} + \dfrac{\text{T.P.}}{\text{P.D.}}x\right) - (\text{P.D.} + x)\left(\dfrac{\text{T.P.} + \dfrac{\text{T.P.}}{\text{P.D.}}x}{365 + \dfrac{\text{T.P.}}{\text{P.D.}}x}\right)}{(\text{P.D.} + x)\left(\dfrac{\text{T.P.} + \dfrac{\text{T.P.}}{\text{P.D.}}x}{365 + \dfrac{\text{T.P.}}{\text{P.D.}}x}\right)}$$

$$= \frac{365 + \dfrac{\text{T.P.}}{\text{P.D.}}x}{\text{P.D.} + x} - 1$$

4 The production process

1 The word *moruna/moruno* may derive from *moro*, meaning Moor or unbaptized (Robe, 1960: 115).

2 The term probably is a contraction of *a casquete quitado*, meaning done in a free manner. *Casquete* refers to helmet or skull-cap and comes from *casco*, meaning skull, hull, casing. The general idea, then, is that of cutting the roots by breaking the 'skull' of the earth. The metaphor of head = earth is used also for other work activities.

3 Beans seeded in April are called a 'pot garden' - raised quickly to eat in the house. A November seeding which is harvested in January is a *huerta de verdanera* (*la huerta* = garden, *verde* = green, *el verano* = summer).

4 The socially expected and sample profit figures are not exactly comparable, since the fields actually tilled were not one hectare in size. The value and surplus calculations are, however, comparable.

5 Sugar-cane production hardly constitutes an exception.

5 Organizing a labour force

1 For comparative descriptions see Erasmus (1956), Firth (1939), Métraux (1951), Smith (1973), Steiner (1957) and Udy (1959). On Panama see Adams (1957), Fuson (1959), Hooper (1945) and Young (1971).

2 The term *junta* also may be applied in a slightly different way to a group of females. During a wake, coffee and cooked food are served to the praying congregation. The small group of women, usually linked by kinship or affinal ties, who help the principal female prepare the food may be said to form a *junta*. No

food or drink is served to this group nor is an obligation to return the work incurred.

3 Since a participant may send a substitute, young men do sometimes work - in place of their fathers.

6 The seeds of a transformation

1 In 1960 in Veraguas 77,914 tons of raw sugar cane were produced from 3,107 hectares, but by 1966, 191,900 tons were being produced on 4,680 hectares (*Censos Nacionales de 1960*, 1963: 65; *Estadistica Panameña*, 1967b: 10).

Epilogue: continuities and changes

1 I am most grateful to the University of Minnesota (Office of International Programs, McMillan Fund, Graduate School) for support rendered.

Bibliography

ADAMS, R. N. (1957), *Cultural Surveys of Panama - Nicaragua - Guatemala - El Salvador - Honduras*, Washington: Pan American Sanitary Bureau.

BENNETT, CHARLES F. (1968). *Human Influences on the Zoogeography of Panama* (Ibero-Americana, 51), Berkeley: University of California Press.

BOEKE, J. H. (1942), *The Structure of Netherlands Indian Economy,* New York: Institute of Pacific Relations.

BOHANNAN, PAUL (1955), 'Some Principles of Exchange and Investment Among the Tiv', *American Anthropologist,* vol. 57, pp. 60-70.

CARLES, RUBÉN (1959), *220 Años del Periodo Colonial en Panamá*, Panama City: Ministerio de Educación.

CASTILLERO C., ALFREDO (1967), *Estructuras Sociales y Económicas de Veragua desde sus Orígenes Históricos, Siglos XVI y XVII*, Panama City: Editora Panamá.

CASTILLERO C., ALFREDO (1970), *La Sociedad Panameña: Historia de su Formacion e Integracion*, Panama City: Direccion General de Planificacion y Administracion de la Presidencia.

CASTILLERO R., ERNESTO J. (1962), *Historia de Panamá* (7th ed.), Panama City: Impresora Panama, S.A.

Censos Nacionales de 1960 (1963), vol. I, 'Produccion Agricola', Panama: Contraloria General de la Republica.

Censos Nacionales de 1960 (1965), vol. III, 'Caracteristicas de las Explotaciones Agropecuarias', Panama: Contraloria General de la Republica.

Censos Nacionales de 1970 (n.d.), vol. I, 'Lugares Poblados de la Republica', Panama: Contraloria General de la Republica.

CHAYANOV, A. V. (1966), *The Theory of Peasant Economy* (eds D. Thorner, B. Kerblay, R. E. F. Smith), Homewood: Richard D. Irwin.

DALTON, GEORGE (1960), 'A Note of Clarification on Economic Surplus', *American Anthropologist*, vol. 62, pp. 483-90.

DIAMOND, S. (1974), *In Search of the Primitive*, New Brunswick: Transaction Books.

DOBB, MAURICE (1973), *Theories of Value and Distribution since Adam Smith* Cambridge University Press.

EPSTEIN, T. SCARLETT (1968), *Capitalism, Primitive and Modern*, Michigan State University Press.

ERASMUS, C. (1956), 'Culture Structure and Process: The Occurrence and Disappearance of Reciprocal Farm Labor', *Southwestern Journal of Anthropology*, vol. 12, pp. 444-69.

Estadistica Panameña (1967a), 'Informacion Agropecuaria', *Serie 'H'*, no. 1.

Estadistica Panameña (1967b), 'Informacion Agropecuaria', *Serie 'H'*, no. 2.

FIRTH, R. (1939), *Primitive Polynesian Economy*, London: Routledge & Kegan Paul.

FIRTH, R. (1963), *Elements of Social Organization*, Boston: Beacon Press.

FIRTH, R. (1975) 'The Sceptical Anthropologist? Social Anthropology and Marxist Views on Society', in *Marxist Analysis and Social Anthropology* (ed. M. Bloch), London: Malaby Press.

FRANK, ANDRÉ GUNDER (1967), *Capitalism and Underdevelopment in Latin America*, New York: Monthly Review Press.

FRANK, ANDRÉ GUNDER (1969), *Latin America: Underdevelopment or Revolution*, New York: Monthly Review Press.

FRANKENBERG, RONALD (1967), 'Economic Anthropology: One Anthropologist's View', in *Themes in Economic Anthropology* (ed. R. Firth), London: Tavistock.

FUSON, R. H. (1958), *The Savanna of Central Panama: A Study in Cultural Geography*, Ann Arbor: University Microfilms.

FUSON, R. H. (1959), 'Communal Labor in Central Panama', *Rural Sociology*, vol. 24, pp. 57-9.

GEERTZ, CLIFFORD (1963), *Peddlers and Princes*, University of Chicago Press.

GODELIER, MAURICE (1972), *Rationality and Irrationality in Economics* (trans. B. Pearce), London: New Left Books.

GUDEMAN, STEPHEN (1976a), *Relationships, Residence and the Individual*, London: Routledge & Kegan Paul.

GUDEMAN, STEPHEN (1976b), 'Saints, Symbols and Ceremonies', *American Ethnologist*, vol. 3, pp. 709-29.

HARRIS, MARVIN (1959), 'The Economy Has No Surplus?', *American Anthropologist*, vol. 61, pp. 185-99.

HOOPER, O. (1945), 'Aspectos de la Vida Social Rural de Panama', *Boletín del Instituto de Investigaciones Sociales y Económicas*, vol. 2, pp. 67-315.

KEYDER, CAGLAR (1975), 'Surplus', *The Journal of Peasant Studies*, vol. 2, pp. 221-4.

LACLAU(H), ERNESTO (1971), 'Feudalism and Capitalism in Latin America', *New Left Review*, no. 67, pp. 19-38.

LOTHROP, SAMUEL (1950), *Archaeology of Southern Veraguas, Panama* (Peabody Museum Memoir, vol. IX, No. 3), Cambridge: Peabody Museum.

McWILLIAMS, C. (1939), *Factories in the Field*, Boston: Little, Brown.

MANDEL, ERNEST (1968), *Marxist Economic Theory,* New York: Monthly Review Press.

MARX, KARL (1967a). *Capital,* vol. I, New York: International Publishers.

MARX, KARL (1967b), *Capital,* vol. III, New York: International Publishers.

MEEK, RONALD, L. (1973), *Studies in the Labour Theory of Value* (2nd ed.), London: Lawrence & Wishart.

MÉTRAUX, A. (1951), *Making of a Living in the Marbial Valley,* Paris: UNESCO.

NAKAJIMA, C. (1969), 'Subsistence and Commercial Family Farms. Some Theoretical Models of Subjective Equilibrium', in *Subsistence Agriculture and Economic Development* (ed. C. R. Wharton), Chicago: Aldine.

NASH, M. (1961), 'The Social Context of Economic Choice in a Small Society', *Man,* no. 219, pp. 186-91.

O'LAUGHLIN, B. (1975), 'Marxist Approaches in Anthropology', in *Annual Review of Anthropology* (ed. B. J. Siegel), Palo Alto: Annual Reviews.

ORTIZ, SUTTI (1973), *Uncertainties in Peasant Farming*, London: Athlone Press.

PEARSON, HARRY W. (1957), 'The Economy Has No Surplus: Critique of a Theory of Development', in *Trade and Market in the Early Empires* (eds K. Polanyi, C. M. Arensberg, and H. W. Pearson), Chicago: Free Press.

Plan de Veraguas (1968), Panama: Obispado de Santiago.

POLANYI, KARL (1968), *Primitive, Archaic, and Modern Economies* (ed. G. Dalton), New York: Doubleday.

RICARDO, DAVID (1951), *The Works and Correspondence of David Ricardo*, vol. I (On the Principles of Political Economy and Taxation), (ed. P. Sraffa), Cambridge University Press.

RICHARDS, A. I., STURROCK, FORD and FORTT, JEAN M. (eds) (1973), *Subsistence to Commercial Farming in Present-Day Buganda*, Cambridge University Press.

ROBBINS, LIONEL (1932), *An Essay on the Nature and Significance of Economic Science*, London: Macmillan.

ROBE, STANLEY L. (1960), *The Spanish of Rural Panama: Major Dialectal Features*, Berkeley: University of California Press.

ROBINSON, JOAN (1962), *Economic Philosophy*, New York: Doubleday.

ROBINSON, JOAN (1965a), *The Accumulation of Capital* (2nd ed.), New York: St Martin's Press.

ROBINSON, JOAN (1965b), 'Piero Sraffa and the Rate of Exploitation', *New Left Review*, no. 31, pp. 28-34.

ROBINSON, JOAN and EATWELL, JOHN (1973), *An Introduction to Modern Economics*, London: McGraw Hill.

ROGERS, EVERETT M. (1969), 'Motivations, Values, and Attitudes of Subsistence Farmers', in *Subsistence Agriculture and Economic Development* (ed. C. R. Wharton), Chicago: Aldine.

SAHLINS, MARSHALL (1972), *Stone Age Economics*, Chicago: Aldine-Atherton.

SALISBURY, R. F. (1962), *From Stone to Steel*, Melbourne University Press.

SAUER, CARL O. (1966), *The Early Spanish Main*, Berkeley: University of California Press.

SEN, AMARTYA K. (1966), 'Peasants and Dualism With or Without Surplus Labor', *The Journal of Political Economy*, vol. LXXIV, no. 5, pp. 425-50.

SMITH, ADAM (1910), *The Wealth of Nations*, vol. I, London: J. M. Dent.

SMITH, CAROL A. (1975), 'Examining Stratification Systems Through Peasant Marketing Arrangements', *Man* (N.S.), vol. 10, no. 1, pp. 95-122.

SMITH, M. G. (1973), 'Patterns of Rural Labour', in *Work and Family Life* (eds L. Comitas and D. Lowenthal), New York: Anchor.

SRAFFA, PIERO (ed.) (1951), *The Works and Correspondence of David Ricardo*, vol I, Cambridge University Press.

SRAFFA, PIERO (1960), *Production of Commodities by Means of Commodities*, Cambridge University Press.

STEINER, F. (1957), 'Towards a Classification of Labour', *Sociologus*, vol. 7, pp. 112-30.

TAYLOR, FREDERICK (1911), *The Principles of Scientific Management*, New York: Harper.

UDY, S. (1959), *Organization of Work*, New Haven: Human Relations Area Files.

WHARTON, CLIFTON R. (1969), 'Subsistence Agriculture: Concepts and Scope', in *Subsistence Agriculture and Economic Development* (ed. C. R. Wharton), Chicago: Aldine.

WHARTON, CLIFTON R. (1971), 'Risk, Uncertainty and the Subsistence Farmer', in *Studies in Economic Anthropology* (ed. G. Dalton), Washington: American Anthropological Association.

WEST, ROBERT C. (1952), *Colonial Placer Mining in Colombia*, Baton Rouge: Louisiana State University.

YOUNG, P. (1971), *NGA WBE: Tradition and Change Among the Western Guaymí of Panama*, Urbana: University of Illinois.

Index